OUTSIDE SHEATHING

CAP

GUARD

STRINGER

BULWARK STANCHIONS

COVERING BOARD

GUARD

SHEAR PLANK

SHELF

CLAMP

CEILING

Y. CEDAR KNEES

BILGE CLAMPS

HATCH COVERS

GUM CAP

COAMING

CARLIN

1" TIE ROD

DECK

COAMING

GUARD

MAIN DECK BEAMS

DIESEL
ENGINE

FUEL OIL TANK

FRAMES

ENGINE BEARERS

1¼ × 6 GUM LINER

SISTER KEELSONS

KEELSON

KEEL

SECTION AT 4

SECTION AT 8

¾ S.W.R.

Box

T.B.

T.B.

CAP 1½ × 8 GUM

APRON 6" Y. CEDAR

Y. CEDAR
B. HOOK

STEM BAR 2×2½ G. IRON
LOWER END TO BE DRAWN OUT
TO ½×8 FASTENED BY ¾ C.S.K.
SPIKES ON 14" CRS

STEM 10×18 GUM

4×10 FIR

MODEL 8-268A
G.M. DIESEL

FRESH WATER TANK
730 IMP. GALLONS

1¼ × S TIG
BETWEEN
ONS 4×6 FIR

GUM S 10"

⅞ G.I. BOLTS 12" CRS

W.T.B.

⅞ G.I. BOLTS EVERY 24

KEEL 11½ S × 13½ M FIR

W.T.B.

INBOARD PROFILE

ELMER
MINOSKY

FISHING
for a living

One of the Canadian Fishing Company's classic wooden drum seiners finishes a salmon set in Johnstone Strait in the late 1980s.

FISHING
for a living

ALAN
HAIG-BROWN

HARBOUR PUBLISHING

Published by
HARBOUR PUBLISHING
PO Box 219,
Madeira Park, BC Canada V0N 2H0

Published with the assistance of the Canada Council and the Government of British Columbia, Cultural Services Branch.

Edited by Daniel Francis
Cover design, page design and composition by Roger Handling/Glassford Design
Cover photograph, *Winter Set*, by Vance Hanna
Back cover photograph by Brian Gauvin

Photograph credits: AHB—Alan Haig-Brown; BG—Brian Gauvin; CI—Commercial Illustrators; CRMA—Campbell River Museum and Archives; FT—Finning Tractor; IPHC—International Pacific Halibut Commission; IW—Image West; JHS—Jewish Historical Society of British Columbia; JW—Jack Wrathall; NFB—National Film Board of Canada; PABC—British Columbia Archives and Records Service; PAC—Public Archives of Canada; UBC—University of British Columbia Library, Special Collections; VCA—City of Vancouver Archives; VMM—Vancouver Maritime Museum; VPL—Vancouver Public Library; WMA—Whatcom Museum and Archives.

Canadian Cataloguing in Publication Data

Haig-Brown, Alan, 1941-
  Fishing for a Living

  ISBN 1-55017-093-7

  1. Fishers—British Columbia. 2. Fisheries—British Columbia. I. Title.
HD8039.F652C243 1993 338.3'727'09711 C93-091647-6

Printed in Canada

For my children
William
James
Helen
Linda

CONTENTS

CONTENTS

The *Denman Isle* on a herring set. (IW)

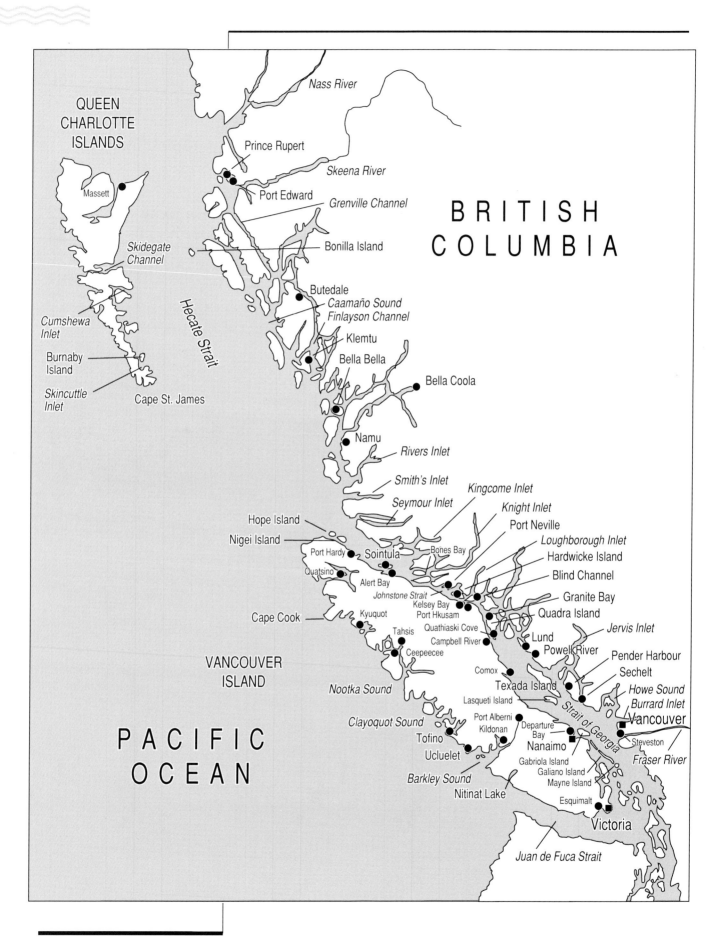

QUEEN
CHARLOTTE
ISLANDS

Massett

Skidegate
Channel

Cumshewa
Inlet

Burnaby
Island

Skincuttle
Inlet

Cape St. James

Hecate Strait

Nass River

Prince Rupert

Skeena River

Port Edward

Grenville Channel

Bonilla Island

Butedale

Caamaño Sound
Finlayson Channel

Klemtu

Bella Bella

Bella Coola

Namu

Rivers Inlet

Smith's Inlet

Seymour Inlet

Kingcome Inlet

Knight Inlet

Port Neville

Loughborough Inlet

Hardwicke Island

Blind Channel

Granite Bay

Quadra Island

Jervis Inlet

Lund

Powell River

Pender Harbour

Sechelt

Howe Sound

Burrard Inlet

Vancouver

Steveston

Fraser River

BRITISH
COLUMBIA

Hope Island

Nigei Island

Port Hardy

Sointula

Quatsino

Alert Bay

Bones Bay

Johnstone Strait

Kelsey Bay
Port Hkusam

Quathiaski Cove

Campbell River

Cape Cook

Kyuquot

Tahsis

Ceepeecee

VANCOUVER
ISLAND

Nootka Sound

Clayoquot Sound

Tofino

Ucluelet

Barkley Sound

Nitinat Lake

Comox

Texada Island

Lasqueti Island

Port Alberni

Kildonan

Departure
Bay

Nanaimo

Gabriola Island

Galiano Island

Mayne Island

Strait of Georgia

Esquimalt

Victoria

Juan de Fuca Strait

PACIFIC

OCEAN

PREFACE

It seemed to me, as a young boy growing up in Campbell River in the 1950s, that there were only two kinds of legitimate work—logging and fishing. I chose fishing. Herb Assu took me onto his 77-foot seiner *San Jose* in the summer of 1960 because I had married his daughter and at eighteen did not have a lot of prospects. Herb was a patient man. So was Jimmy Mitchell, the deck boss. They didn't yell at me much, but they gave me a lot of opportunities to learn. When we finished the dog salmon season down in Satellite Channel in the November fog, I went to work cutting down old telegraph poles on an abandoned logging railway grade. I thought I was going to make fence posts out of the cedar and sell them. Truth is, I didn't know what I was doing, so when I got word that Herb wanted me to go out winter herring fishing with him, I ran to pack my gear.

Those were the days when we got paid $8.80 a ton for herring that a couple of decades later was fetching $2500 a ton. I was the only one on the boat who wasn't a seine skipper during salmon season. I was also the only non-native.

I learned a lot that winter. Among other things, I learned about learning, so that eventually I returned to school to finish grade twelve and then went on to university to train as a teacher. Fishing paid the way, and the people with whom I fished kept me focussed. They told me about the

James Walkus's fibreglass seiner *Pacific Joye* was built at Mike Goldrup's Campbell River shipyard in 1978. Here she brings a small bag of salmon over the stern during an opening off the mouth of the Fraser River in September of 1988. (AHB)

The salmon fleet puts out from Bones Bay for a 6:00 p.m. Sunday opening in August 1945. At times like this the boats "race" to see which are faster, although they all average a top speed around nine knots.
(John Mailer/PAC PA-145355)

residential schools and about the Japanese Canadians who had built and lost the boats which now carried me safely through southeasters off Cape Mudge. On the weekends I visited retired skippers like Jimmy Hovell, who told me about the beginnings of the commercial fishery and the times before that when the Kwagul'th ruled Johnstone Strait. My deck boss on that winter of herring, Tony Roberts, made it clear that they still did.

After graduating from university and spending a year teaching, I decided it was time to quit fishing. My wife and I drove to Expo 67 in Montreal. By early August I was back on the dock at Campbell River waving goodbye as my old boat left for the fishing grounds. But Herb knew better. "Go home and get your slicker gear and get on board," he told me. So I did. They say sockeye fever is incurable.

Even when I finally stopped going fishing in the 1970s and was living in the Cariboo, I found myself down on the docks as soon as I got back to the coast. I saw many of the wooden boats I had known disappear, and grumbled as aluminum and steel boats replaced them. I became a boat bum with a mission. I would record all of this for posterity.

My old skipper, Herb Assu, had died, but the legendary Charlie Clarke had only recently retired as Nelson Brothers' head skipper when I went to see him in 1986. I had never met Charlie but I had fished for Nelson Brothers and knew he was a revered figure in the fleet. It was with Charlie that this book began. He sent me on to Louie Percich and Dick Anzulovich, who sent me on to others. I moved to Vancouver just as a young entrepreneur from Ucluelet, Leon LaCouvee, was starting a magazine, the *Westcoast Fisherman*. I became editor and was able to turn my passion into a job, walking the docks and talking to the people who designed, built and fished the boats.

This book is the result of a three-decade infatuation with fishing boats and fishing people. In spite of an ugly underbelly of racism and resource abuse, the industry has produced some beautiful boats and has rewarded its people with lives filled with dignity. I am proud to have shared in the experience and humbled by the stories that have been given to me to tell.

A note on terminology. The Women's Maritime Association has formed in the USA to help educate women about the opportunities at sea and to give women some survival skills in what has been a male-dominated world. The organization draws its membership from women working in the maritime industries, from supertankers to one-person trollers. In 1992 I asked then-president Anne Mosness about the use of the term "fisherman."

The *May S* was built at Kishi Brothers Boatyard in 1927. Here she seines salmon in Johnstone Strait in 1989. (IW)

"We aren't fond of it," she replied, "but we are much more concerned about the issues of sexual abuse and harassment at sea and with gaining equal opportunity for our members."

I suggested that the word "fisherman" carried a lot of the history of the English language in it. Even if women had been fishing for years, the term demonstrated that this had not been the social norm. Since I loved the language I appreciated the cultural load of the word.

I had gone too far. Anne explained politely that she quite understood my sentiments. She, too, "would love a language that revolved around her gender."

I felt quite properly chastised, and in recognition of Anne's simple logic I have used the gender-free word "fisher" throughout this book.

It is difficult to thank properly all the people who contributed to this book. There are so many of them. Much of the text is based on interviews with Robert Allen, Dick Anzulovich, Edgar Arnet, Harry Assu, Hisao Atagi, Bert Benson, Charlie Clarke, Florence Davidson, Al Dawson, Burton Drody, O.C. "Fergy" Ferguson, Ken and Vickie Hagen, Jean

(Hideko) Hamagami, Ken Hamagami, Ed Hanson, Henry Helin, "Hutch" Hunt, Neil Jensen, Jim Kishi, Fred Kohse, Hank McBride, Mamoru Madokoro, Richard Martinolich, Sam Matsumoto, Bert Menchions, Ray Michaelson, Shin Nakade, Ted Nakatsu, Scotty Neish, Haakon Novik, Louie Percich, Al Renke, Brad Scott, Roger Skidmore, Randy Thompson and Bill Wilson. The list could be many times longer and still be incomplete. I wish to thank everyone who shared the stories which have contributed to my understanding of the fishing industry as it is, and as it was.

Within the publishing world, I have to acknowledge the help of Brad Matsen of the *National Fisherman*, who has taken a great deal of time over the years to coax coherent, concise sentences from me. At Harbour Publishing, Pat Sloan and the editorial team of Dan Francis and Mary Schendlinger have exercised their craft to my benefit. Roger Handling has my appreciation for his design work in presenting all of the bits and pieces in a coherent format.

This book does not pretend to be a definitive overview of the industry. It is simply my look at an incredibly diverse community. I take full responsibility for any errors of omission and for my interpretations of the history which may differ from those of my informants.

THE SET

The bulk of the big green-hulled seine boat moves at slow speed through the January darkness riding the groundswell of Hecate Strait. The boat's running lights glow red and green on the dodger that circles the flying bridge just ahead of the wheelhouse on the cabin top. Deck lights cast a warm pool of yellow over the working deck and the great mound of net piled on the stern. Riding in the wash of the big boat, snubbed tightly bow to stern, a 20-foot power skiff rises and falls on the swells in mimicry of the lead boat. Two men are in the skiff. One stands alert amidship ready to throw the clutch on the skiff's engine into reverse and pull the net from the seiner's stern; the other is in the bow ready to retrieve the painter when it is released on a signal from the seiner.

On the deck of the big seiner other men stand at their positions: one to release the skiff, one to control the release of the cable purse line. The deck boss, off to the side, runs his eyes over the rigging, checking once more that each block and line is in place to set the net. On the bridge, eyes focus on the paper spooling from the green metal box of the Echo-lite echo sounder. Down one side of the paper the steady click of the needle traces the bottom of Hecate Strait, twenty-five fathoms under the keel of the boat, which forms a line down the other side of the paper. Between the two lines a darkening grey area signifies the mass of herring that has begun to materialize with the darkness.

The *Adriatic Sea* with the winter herring fleet in 1943. She was built at Stanley Park Shipyard in 1940 for Sam Jasich. (R. Wright/PAC PA-145364)

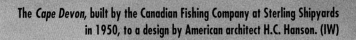

The *Cape Devon,* built by the Canadian Fishing Company at Sterling Shipyards in 1950, to a design by American architect H.C. Hanson. (IW)

Tony Anzulovich skippered the *Northland* on winter herring in 1943. The crew is shown putting a rope strap around the heavy cotton net. The powered hook on a single fall will then be lowered and another "fleet" of net will be pulled aboard. The introduction of the hydraulic power block a decade later simplified this operation considerably.
(R.Wright/PAC PA-145356)

"Looks good. About 500 tons in a set," says the Captain to the man at his elbow. "Go tell the boys to stand by."

Eight men and hundreds of fathoms of lines, cable and web are ready. In the engine room the big diesel idles ahead. The skipper maps the size of the school, checks the lights of the boats cruising around him. The short toot of a horn carries across the water from a boat a few hundred yards to the west. "A little too soon and a little too far up on the edge," the Captain muses to himself. But he knows that now other boats will be prompted to follow and very soon there will be nets out all over and the school will begin to break up. It is time to go.

He reaches through the open wheelhouse window and pulls the

THE SET

whistle cord. A short blast pierces the night and releases the tension on deck. A quick flip of a wrist and the skiff's painter is free of the restraining cleat. The skiff man's foot pushes the clutch lever to the reverse position as his hand depresses the throttle. The man in the bow of the skiff pulls in the painter as the skiff turns on the end of the net. He spins the tow line off the bow post and runs it around to the aft towing bollard as the skiff turns away from the net to begin towing.

On board the seine boat the purse line snakes out through the snatch block hanging from a davit on the side of the boat. It shoots aft passing through the row of brass purse rings that hang on the side of the seine table. The leads on the lead line send out a steady drone as they pass over the stern. Each ring gives a dull thump as it leaves the wooden hull for the cold water. The mound of fine herring web mists over the stern to disappear in the black water as the rows of corks settle to float on the surface marking the circling course of the boat.

The Captain turns the big boat back toward the skiff in a circle scribed to close its perimeter just after the last of the net leaves the stern. The man in the bow of the power skiff waits until the groundswell lifts him nearly level with the plunging bow of the seine boat, then throws the heaving line to the waiting deckhand. The end of the net is freed from the power skiff as the powerful deck winch begins to winch it in. The deck crew works in a steady frenzy to secure and release lines in harmony, which soon has the cable purse line turning over the twin winch heads and onto the spools. The bottom of the net purses under the swarming herring. The power skiff takes up position on the off side of the boat towing a bridle fixed near the bow and stern of the parent boat.

Brailing herring in 1943 (top) and February 1961 (left) took great teamwork. A man on the winch lowered the big dip net, or brailer, into the massed herring while an operator on the handle, usually the deck boss, pulled it toward the back of the boat. A third person, also at the winch, worked the skimmer line which pulled the brailer through the fish. At a signal from the person on the handle, the first man winched the brailer with its two-ton load out of the water. A fourth person pulled or winched the vang, a small pulley line running from the China boom near the top of the cabin to a point near the end of the main boom. It swung the main boom and brailer over the hatch. A fifth man sometimes worked the vang on the other side to swing the brailer back. Finally a sixth person, the asshole man, usually the most junior member of the crew, pulled a small line to a trip block on the brailer that dumped the fish out through the bottom into the hold of the seiner or a packer tied alongside. If the asshole man miscalculated, the herring all dumped on deck. That was the way it was done on the *San Jose* where I worked in 1960–61. (R. Wright/PAC PA-145353; AHB)

The Purse Seine Set

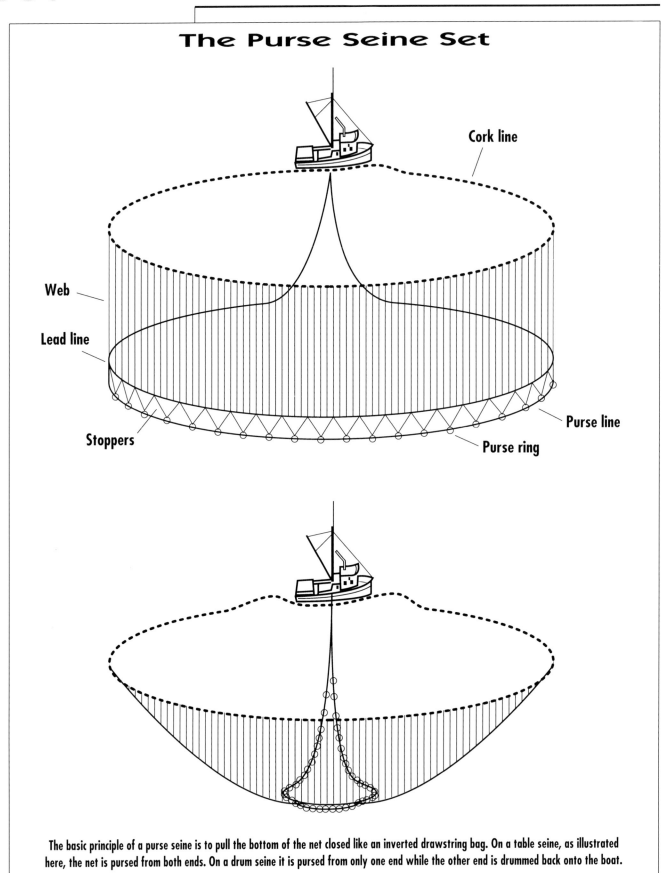

Cork line

Web

Lead line

Stoppers

Purse line

Purse ring

The basic principle of a purse seine is to pull the bottom of the net closed like an inverted drawstring bag. On a table seine, as illustrated here, the net is pursed from both ends. On a drum seine it is pursed from only one end while the other end is drummed back onto the boat.

THE SET

When the rings emerge from the depths at the side of the boat to signal that the bottom of the net has been pursed, a light alerts the power skiff to ease off. The bottom of the net is hoisted aboard and the net is hauled back through a massive power block run up on the end of the boom to hang suspended over the seine table. Two men struggle to pile the mass of heavy web while others coil the attached cork and lead lines. A little more than an hour after the horn has signalled the start of the set, most of the net has been returned to the boat. Now a big bag floats on the black water alongside the seiner. "Five hundred tons at least," estimates the Captain from his position on top of the cabin. "The *Western Warrior* is coming on our cork line to take a load."

Soon a second seiner has lifted the floating cork line and is swinging a powered dip net into the mass of tightly packed herring. The first boat does the same. As the brailer net hits the water, it is hauled through the mass of fish by a skimmer line, then lifted clear of the water to bear its two-ton load of herring, in time with the roll of the groundswell, over the open hatch of the seiner. A trip block is let go to open the bottom of the brailer as it passes over the open hatch. Six men working in harmony with the groundswell repeat the operation until 120 to 150 tons of herring have been brailed onto each boat. Another boat takes up position on the cork line before the net is emptied and the bunt end is hauled aboard.

By the time the skiff has been re-secured astern, and the decks washed down, the first light of day is softening the sky over the mainland mountains. A thin dark wedge on the horizon marks the Queen Charlotte Islands off to the west. After ten hours in the wheelhouse, the Captain climbs down to the galley for a hot breakfast. He feels good. This is what he does and he does it well. He has been setting and hauling seine nets since the gear was brought up from the US before World War One. Now he relaxes on a good big boat riding easily in the groundswells. One more set has been completed with a good catch brought aboard by a good crew working as a single-minded team.

The skiff can be seen at the top of this illustration, waiting for the seiner to finish setting the net. As the net leaves the stern of the seiner, the lead line pulls the bottom down until it hangs like a sail in the ocean currents. (Western Fisheries, 1971)

Below, the *Western Monarch* brails herring onto a packer from another boat's set. The canvas skirt hanging over the side of the boat protects the net from damage by the brailer. Note the floats tied onto the lead line to prevent it sinking into the bottom mud. (R.Wright/PAC PA-145358)

A RECORD SET

Tony Smith and Hans Zimmerman dry up the fish by taking in the extra web on a herring set by the *Northern Dawn* in 1988. (BG)

On the 15th of March, 1987, Don Dawson brought his steel seiner *Snow Cloud* alongside his partner John Lenic's *Ocean Marauder* at the Ocean Fisheries plant on the Vancouver waterfront. Word of his 970-ton roe herring set in Barkley Sound had preceded him, as had most of the fish from the set. Payments on such a set are complicated by various bonus arrangements and not even other fishers will know exactly how much money changes hands, but the best guess at the time was that something around $2500 per ton would have been paid for these fish. Company officials were so impressed that they had a case of champagne, complete with glasses, waiting to welcome the exhausted skipper and crew. After the *Vancouver Sun* photographer had left and some of the hoopla had subsided, I sat in the galley with Don Dawson while he told the well-known dragger skipper Vigo Mark the story of the $2.5 million set, likely the most valuable single set in the whole BC fishery.

"We left Vancouver on March 1st and arrived in Barkley Sound on the 2nd. We laid up in Port Alberni for four or five days. It was decided that if there were adequate stocks on the morning of the 12th we would have an opening. The Department of Fisheries was concerned because there had been a series of storms on the West Coast and with a big swell running into Barkley Sound there can be a problem with some of the smaller boats handling the fish. On the test fishing the herring were pulling the corks down, so we knew that these were heavy fish to handle. They seemed to have a tendency to sound in the net and that puts more pressure on the net.

"On the morning of the 12th they got me and three other seiners, the two test vessels and the patrol boat to work a grid to assess the stock. We started at 4:00 a.m. and by 10:00 they were able to say that there would be an opening announced for twelve noon.

"We set on a good school right off the bat, but another boat set inside our net. We were in very shallow water and couldn't drum our net back or anything. We had to wait until the other boat got his net back and by that time our lead line was well

This 1989 herring opening in Barkley Sound is typical of the congestion as dozens of boats try to set their nets on a single body of fish. Boats commonly collide, or run over each other's nets. (BG)

anchored in the mud. So we ripped up [our web]. When we brought up our [purse] rings, the mud was just caked in there. There was a 30- to 40-fathom hole in the net, with a tail [of web] in it, about two strips of web above the lead line, so we had to lace that up as we brought the net aboard.

"By the time we got the net aboard, it was about ten to two and all the other boats were on sets. A 'fish-cop' came on the radiophone and said that they were 1500 tons short of the quota, so the obvious thing to do was to see where the hell these fish were. The tide had started to ebb, so I figured that they would head out and they would go through Macoah Passage, out through David Islands. I went out that way and started looking [with the sonar] and we found them in Macoah Passage. At first I didn't know what was going on because the edge comes up very quickly there, from 19 fathoms to 5 fathoms, and then there is a flat in by the river. The sonar didn't seem to show me any edge on long range so I turned and went over there. When I got there I shortened down the range on the sonar. At 400 feet the picture changed. There was no bottom, it was solid fish from top to bottom. I was going quite quickly trying to find some fish. By the time I could turn the boat around up there on the flats I knew there were a lot of fish.

"The way the fish were lying against the edge called for a left-hand [turning] set in order to lead the fish in the direction they were moving. The way I was positioned over the flat I saw there was so damn much fish there it was hardly worth fooling around. Just get her out! I'd already told the crew to stand by, so I let the skiff go on the shallow and made a right-hand set running right along the outside of the edge. I let the net go in 5 fathoms of water, then ran [the boat] out over the edge, ran along the outside of the edge, and came back to my skiff in 5 fathoms of water with the rest of the net. If I'd had time to make a left-hand set I think we would have got 1500 tons.

When an opening is announced over the radiophone by the Department of Fisheries, the boats race to find the fish. Once on the fish, some begin setting while others may look for a better concentration of fish with their sonar. (IW)

"We drummed back about a quarter of the net and got the rings up. We knew that we had a lot of fish so I called for John Lenic on the *Ocean Marauder* to come [and tie his boat] on the cork line. He was still on a set so we had to wait about forty-five minutes. While we were waiting we got everything ready. We got our breast well strapped up, picked up a fleet five rings off the front end of the net, brought the purse rings up in the air. Then we got the bunt web and put a heavy strap on that and put it on the double fall.

"When the *Ocean Marauder* was in position on the cork line we started drumming [the net back]. Then as we closed the fish off, the corks got

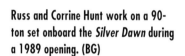

Russ and Corrine Hunt work on a 90-ton set onboard the *Silver Dawn* during a 1989 opening. (BG)

heavier and heavier. It was all we could do to drum it and then the corks started to go down. But I have eighteen bladders [big floats] on the cork line and that helped like a skiff being on there. A few fish did go out. We kept the drum going until it stalled. We were past the halfway mark on the net. By that time we were in a position to put the *Marauder*'s power skiff on the cork line between his bow and our stern. Our power skiff was towing full bore to keep everything up.

"The *Ocean Marauder* was already loaded from the set that John had made, but he took another 20 tons. We had started to list from the weight of the fish pulling down the net, so we pumped the water out of our starboard tanks. Then we pumped ourselves full of herring. The fish

When a boat has pursed its net, a packer or another seiner will come alongside to help pump out the fish, which will often fill more than one boat. Power skiffs tow the two boats apart while they pump the fish out of the net. (IW)

were still swimming around so it wasn't too hard to pump.

"After that, the first packer we could get, the *Eastward Ho*, only had room for 50 tons. They told us there were no more packers for four hours so we cinched everything down and decided to get some sleep. It was 2:00 a.m. of the 13th and I'd been up since 4:00 a.m. on the 12th. We had set at about 2:20 p.m. on the 12th. We kept a couple of guys on watch. After an hour one of the guys called me to say that the net was tearing, but we put a strap on it and closed it up. By this time the net was on the bottom, but there was a swell running and I was worried that the web would split.

"About that time the *Ocean Horizon* arrived on the scene. He towed us, the net, the fish, and the *Marauder* with our bridle a couple of miles over behind Stopper Islands. The *Marauder* dropped her anchor and held up there in 14 fathoms. That's where we stayed and put our fish piecemeal onto other boats as they came along.

"By the morning of the 13th the fish were pretty well all dead and lying on the bottom. We never really dried up [the net], we just pumped the fish off the bottom. The sounder showed 86 feet to the bottom and when we put a sounding lead down to the fish we measured 72 feet, so there was 14 feet of dead fish lying in the net on the bottom. It was very hard to pump out because the fish were so dead and packed, just like concrete. The *Ocean Marauder* did all the pumping. She can normally pump up to 200 tons per hour, but with the dead fish it was taking three

The process of drying up the net so that the web comes up evenly under the fish can be backbreaking work, but if there is too much water in the net it is difficult to pump the fish. Here Con Charleston directs the operation on the *Princess Colleen* in 1989. The hydraulic lines for the pump enter the net just to the left of centre but the main hose at the right top is still flat and empty of fish. (BG)

A RECORD SET

The crew of the *Snow Cloud* toast their record set, March 1987. Skipper Don Dawson is second from right. (AHB)

hours for 80 tons. The weight was so great that the brake on the seine drum couldn't hold, so we had to put a cinch line on the net at the stern and tie it down. A couple of times we tried to use the drum to roll the fish in the net but it wasn't until just before noon on the 14th that we could move it. By that time, we had about 200 tons left.

"We put fish on thirteen different boats, including ourselves. On the morning of the 14th George Brown on the *Pacific Harvester* took 150 tons and that was a big help. The *Ocean Ranger* took the last 3 or 4 tons.

"It was quite a set, fifty-four hours from start to finish. It was really a partnership with John Lenic on the *Ocean Marauder*, and the other boats helped a lot too. You don't do a set like that alone."

The *Invercan IV*, built in New Westminster in 1945, works on a set in Johnstone Strait, while the other boat (left) prepares to bring its fish aboard. (IW)

FISHING BOATS, FISHING PEOPLE

"You could walk across the river on the backs of the salmon." This oft-repeated image describes the bountiful harvest of fish that the first Europeans found in the waters of the Pacific Northwest. As salmon move onto their birthing gravel to lay the eggs that will renew the life cycle, it does almost seem that one can walk on them. It was this plenty that helped the First Nations people inhabiting the coast and river drainages to build their complex and diverse cultures thousands of years before the arrival of the Europeans.

This wealth of salmon and other marine life was nurtured in the enfolding contours of the coastal geography. Herring, eulachan, cod, mussel, clam, crab, seaweed, sea urchin and pilchard were plentiful and accessible. The inlets and islands of the coast provide many miles of rich intertidal zone in which the bounty was harvested. The mountains behind the coast caught the rains and turned them back to the sea in the form of rivers with rich, feed-filled estuaries. The same rains nourished dense coniferous forests, providing wood for canoes and fish traps, as well as roots and bark from which to fashion fishing gear.

The traps and other fishing technologies of the Native peoples

A rowed seine boat in the American San Juan Islands in 1905. The oarlock on the bow is for steering when the net is being set. The San Juan Island location indicates that this is probably an early purse seine intercepting Fraser River-bound salmon. Note the sunken three-masted sailing ship in the background. **(WMA)**

were designed to harvest only what was needed while allowing adequate escapement for the peoples living upriver and for the spawning grounds. The introduction of a cash economy by the Europeans changed all that. Fish were not the first cash crop on the coast, of course. Sea otter pelts were traded from the time of Captain Cook, skinned and preserved for the long trip to the markets across the Pacific. When the sea otter were gone, river gold became the new bonanza. And then the salmon.

For the European immigrants of the nineteenth century, salmon represented a convenient food. Purchased from the First Nations people, it was either fresh, smoked or dried. The Hudson's Bay Company began salting salmon at Fort Langley on the Fraser for export to Hawaii around the middle of the century. In the 1870s the first successful salmon canneries were established along the lower Fraser River, turning out a product that could survive the long voyage around the Horn to England. Traditional native fishing methods couldn't keep up with the demand for fish, and fishers using gillnets began working the lower reaches of the Fraser. This type of net hangs like a curtain in the water. The individual meshes of the net are made to a size that allows a fish's head to pass through but not its body. Once the fish is in the net, its gills prevent it from backing out.

Trolling for salmon with metal hooks on lines has been practised since iron first appeared on the coast. Commercial troll fishers concentrated their efforts on coho and chinook (spring) salmon and marketed their high-quality catch to the fresh fish market. The development of powered packers that could ice the fish and transport them to urban markets

By 1910 the conversion of the old rowed and sailed Columbia River-type gillnet skiffs to gasoline power was well underway on the Fraser River. There were a number of local manufacturers of the little one-cylinder engines but the Easthope Brothers' engine became the best known. The pile of net on the stern of the powered gillnetter was pulled by hand until the introduction of the powered drum in the 1930s. The letters IMP on the sail indicated that this gillnet boat belonged to the Imperial Cannery on Steveston's cannery row. It was the last of the Steveston canneries when it shut down in 1992. (VPL 2041)

dramatically increased the range and viability of the troll fleet through the 1920s and 1930s, and the advent of power gurdies to retrieve the lines made possible greater production and fishing at greater depths.

Seine boats are the largest and most expensive of the salmon and herring boats. Early capitalists found them attractive because they represented the industrialization of the fishery. Often declared the most efficient fishing craft in the world, seine boats are effective in catching fish that travel in schools at relatively shallow depths. On the east coast of the US seining is used for menhaden, a kind of herring; on the oceans of the world seiners fish for tuna. It is the tuna fishery which has done so much to damage the seiners' reputation because many thousands of porpoise are killed in the process.

In BC seine boats have been used to fish for salmon, herring and, prior to their disappearance, pilchard. By the first years of the twentieth century all the major rivers, from the Fraser in the south to the Nass in the north, had gillnet-supplied salmon canneries. A number of these rivers also had drag-seine operations. In this fishery a large net is set out from the beach around fish schooling at the river mouth. The bottom of the net lies on the bottom so that when the two ends of the net are dragged back ashore the fish are hauled up with it. In most cases these drag seines were controlled by native Indian fishers. The advent of the seine boat made it possible to move out to deeper waters to intercept the river-bound salmon, before they reached the native controlled drag seines. At the same time lucrative pilchard and herring fisheries were being developed with seine boats.

The *Leona H II* was built at Sather Boat Works in Queensborough in 1968. When this photo was taken in 1987, she was owned by Miteru Higo who rigged her for both gillnetting and trolling as indicated by the net drum on the stern, pullies for the trolling lines on the davits just behind the drum and the trolling poles which are let down when the boat is fishing. (AHB)

Between about 1910 and 1920 the seine boat became established as a major force in the coastal fishery. Around 50 feet in length, these first boats had powerful gas engines, for their day, which propelled the boat and helped haul in the gear. The boats caught thousands of salmon or hundreds of tons of herring or pilchard in a single set of the net. They were very versatile, useful for towing out a string of sail-powered gillnet boats and bringing their catch back to the cannery. They could be rigged for trawl fishing, in which a smaller net is dragged along the bottom for cod and sole, or they could be rigged to longline for halibut. Their versatility and high productivity made them very popular with cannery owners.

Fishing is a tough and a dangerous job, but it has always paid better than most and it can be done without any schooling or even the ability to speak English. It is the ideal job for an ambitious immigrant who wants to "get ahead" in a new country. The combination of immigrants with limited command of the ways of the country, natives with limited legal rights, and an expensive piece of equipment like a fish boat provided fertile ground for the capitalist.

The BC fishing industry spawned a number of capitalists in the usual proportion of scoundrels, visionaries, hard workers and opportunists, but all of them faced challenges unique to the industry. Unlike a cannery, where because it is fixed in one place, the usual hierarchy of owner, manager, floor boss and worker can be maintained, the seine boat with a skipper and crew does not yield naturally to the traditional power structure. Seine skippers are not trained in the paramilitary tradition of merchant mariners. Their fishing skills are more important than their navigational skills and less easily measured. An expensive seine boat can be well navigated and carefully maintained, but catch no fish. The good fisher in this setting quickly emerges as part artist, part co-adventurer. After attempting to own all the seine boats, most fishing companies recognized this. There was competition for good skippers and the companies

did much to maintain the allegiance of their good fishers, including large interest-free or low-interest loans for new boats. The major companies, like BC Packers, Canadian Fish, Anglo British Columbia, Queen Charlotte Fish, J.H. Todd, and F. Millerd & Co. Ltd. (fish canners), nurtured close friendships with their skippers. Right up into the 1960s it was routine for company representatives to attend family weddings and funerals.

Fishing companies used native Indian skippers because of their comprehensive knowledge of local tides and fish. Until 1938 Indians were also used to break strikes. Racist attitudes throughout the province had always kept them separate from European and Japanese Canadian fishers, and they didn't see the advantages of common cause. Because Indian fishers had a tradition of taking only as much fish as they needed, the companies were often perplexed by their seeming lack of interest in striving for more fish and bigger boats. The companies engaged in pater-

When Jim Warnock and his dad decided to build a seiner back in 1946 they took advantage of the easy access that ordinary people had to the resources of the land. Jim's dad went up to the head of Jervis Inlet and bought six big yellow cedar logs from a logger. He towed the logs down to his home at Pender Harbour and sawed them in his own sawmill. Jim and his dad took the lumber down to Tom Taylor, who had a boat yard on the North Arm of the Fraser at the foot of Victoria Drive, and Tom built them the *Keranina*. (AHB)

nalistic strategies of "looking after their money for them" and seeing that Indians got the smallest and oldest boats in the fleet. Debt loads were maintained on the books so that a native skipper could not leave and go to work for another company with any ease.

Throughout the pre-World War Two era the Japanese Canadian community played an important role in the industry, as both boatbuilders and fishers. Denied citizenship and direct participation in the lucrative salmon seine fishery, they developed boats and fishing techniques in marginal fisheries, such as the Gulf of Georgia herring fishery, in which no one else was interested. Finally, in 1942, in a paroxysm of racist excess, their fellow Canadians took away their boats, houses and businesses and interned them simply because of their racial origin.

For the aggressive young European immigrant skipper, things were different. Typically he was introduced to the company by a relative who first employed him as a deckhand. He then took out a company-owned boat for a couple of seasons before getting the opportunity to buy one of the older company fleet. If he proved himself on this boat, he received an interest-free or low-interest loan to have a new boat built. By the time that loan was paid off it was time for another new boat. The old boat was bought by the company or resold to a younger skipper.

Built in 1926 by Nakade of Steveston for Kitoshi Tanaka, the *Porlier Pass* is representative of the boats built to meet the high demand during the Roaring Twenties. It was a time when big profits from the rum-running trade and successful fishing ventures fuelled the demand for the still relatively new seine boat. (FT/CI 1816-1)

In this way of doing things, the boatbuilders got lots of business, the fishing company increased its catch capacity, and the fisher got the satisfaction and increased earning potential of a new boat. For government managers, it meant imposing shorter openings against the protests of fishers who had to meet ever larger boat payments. More important, for the field-level managers it meant allowing fisheries that often did not fit well with their perception of the best interests of the fish, but that were mandated by the Canadian political tradition of catering to capital investment at the expense of the resource. Senior members of fishing companies are regularly represented on government advisory bodies, and managers move back and forth constantly between industry and government. Many fishermen look back fondly to the 1950s when Jimmy Sinclair, then Minister of Fisheries, listened to their complaints. This was the period in which the fleets grew far beyond the ability of the resource to support them, resulting in the closure of the herring fishery in the 1960s and the dramatic shortening of fishing times down to the present. A larger, more efficient fishing fleet, combined with continued habitat depredation, prompted the cry: "Too many boats chasing too few fish."

The government responded during the 1970s with the very expensive buy-back program. Considered by many to have been designed to serve the corporate agenda of capital management and the government agenda of people management, the program retired many boats. The fishers who lost their jobs as a result have not shared in the increased value and earning power of the limited-entrance licences. The two people who probably own or control the largest number of fishing-vessel licences in BC today are a businessman with roots in a car dealership and the scion of a wealthy eastern family.

A Vietnamese immigrant fisher recently explained his caution in dealing with the Caucasian Canadian majority by recounting the history of British Columbia as he had learned it from his fellow fishermen. "First they came here and the Indians helped them," he said. "But they took nearly all of the Indians' fish and land, then they took the Indians' children and put them in residential schools and even took their language. After they had broken the Indians down they needed more fishermen so they brought the Japanese. When the Japanese became successful the Caucasians wanted their boats and property so they took it all away. Even when the Japanese were allowed back on the coast after the war, one fisherman had *Fuck off Jap!* written in shit on the window of his boat, and he was born in Canada. Now we are becoming successful and we hear the Caucasians talking about us and looking at our boats. But they won't be able to do to us what they have done to others!"

The stories from our recent past, stories of success and loss, stories told by people still living or only recently gone, are much more than simple curiosities. It is these stories to which we refer to shape our future. A person steering a seine boat will often check the steadiness of the boat's course by looking back at its wake fading off into the distance. By looking at the recent history of this province's fishing industry and the people who have built it, we should be able to better chart the coming decades.

Built by Kishi Brothers of Steveston in 1939 for Henry Jacobson of Finn Slough, the *Eva* was one of the few Easthope-powered gillnetters to fish the 1993 sockeye run on the Fraser River. Her present owner, Henry's nephew Gus Jacobson, has a new fibreglass boat but for sentimental reasons, he keeps the *Eva* in immaculate condition at Finn Slough. (AHB)

The Seine Boat Boom

By 1927 seiners were being built at an incredible rate. A report in Harbour and Shipping *magazine from April of that year:*

"The development of special fishing craft in connection with seining operations on the British Columbia coast has been an outstanding feature of the last two years.

". . . Last year there were about 55 new seine boats built on Burrard Inlet and the Fraser River.

"As seine boats built for the salmon industry are also suitable for seining herring and pilchard, and can be adapted for halibut fishing, besides being useful for towing and freighting, there has been a good demand for this class of boat, and most of the local shipyards are busy on these craft again this spring.

In 1928, BC Packers wanted new seine boats so badly that they had five built to the same set of plans at two different yards. The *BCP No.41* and *BCP No.42* were built at their own Celtic Shipyard, above, on the North Arm of the Fraser, while the *BCP No. 43* through *BCP No. 45* were built at the Burrard Shipyard in Vancouver's Coal Harbour. The *BCP No. 45* gained fame years later when she appeared on the back of the Canadian five-dollar bill. In this picture the *BCP No. 41* and *BCP No. 42* are being readied for launching. The *BCP No. 42* was converted to a yacht by marine diesel manufacturer Will Vivian and became the *Miss Vivian*. She then entered the Forest Service as the *Tamarack*, and later was sold to the Attorney General's Department and was named *Freedom Found*. Today she is the *Tamarack IV*, a pleasure boat in Vancouver. (VPL)

"Seine boats range from 40 to 75 feet in length, with plenty of beam, the larger ones having a breadth of 15 to 17 feet. They are usually built with a fairly wide stern to carry the seine net and large revolving seine table with its roller the full width of the boat. These boats also require large carrying capacity, for a good haul of fish will often run into many tons. One 65-foot seine boat was reporting taking 22,000 humpback salmon at one set with her seine last season, and this load put her down so deep in the water that her deck was practically awash. With humpbacks averaging close to 5 pounds apiece she must have had about 50 tons on board.

"These fishing craft are heavily constructed of fir, with steam bent oak frames. Formerly they were all equipped with heavy-duty gasoline engines, but in recent years oil engines . . . are finding favour . . . .

"The 70-foot seine boat *Western Chief*, built for Nelson Bros. of New Westminster, was equipped with a 90-hp Washington Estep diesel engine; and several 75-foot seine boats for the Canadian Packing Co. operating on the west coast of Vancouver Island had 100-hp Atlas–Imperial diesel engines. These boats all have power winches run off the main engine, and in some cases power for the seine roller also . . . .

"Japanese boatbuilders at both Vancouver and Steveston are also turning out a number of seine boats. Some of these are being equipped with Thorneycroft 36-hp reduction geared gasoline engines, and several others building at Steveston for Japs [sic] and Indians will have 35-hp Gardner oil engines, the boats being about 45 feet long . . . .

"This period of frenetic building resulted in a great many very fine boats, a good number of them still working today.

CHARLIE CLARKE: ONE OF THE FIRST

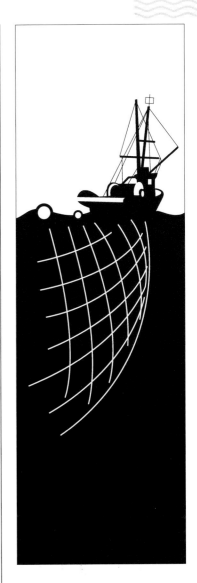

Charlie Clarke was not born to a fishing family but he became one of BC's most successful seine skippers and, eventually, head skipper for Nelson Brothers Fisheries. Born in Washington State in 1897, he grew up in Port Alberni. His father was an Englishman who came up from the States to prospect for minerals on the west coast of Vancouver Island. Like most coastal kids of the era, Charlie started fishing spring salmon out of a dugout canoe for pocket money. The school in Port Alberni only went up to grade four. Once he "graduated" he went to work in the Waterhouse and Green store in Alberni for 25 dollars a month. This led to an offer to go down to Barkley Sound to be assistant storekeeper at the Wallace Brothers' Kildonan Cannery, at double the salary. It was at Kildonan, before World War One, that Charlie saw his first seine boat.

As Charlie recalled it years later, it was kind of an experimental arrangement: "They had a steam tug with a native skipper named Bob Coutes. It came up Barkley Sound with a seine net on a scow alongside. They set their net at daylight on a Saturday. It was dark before they finished the set and they had filled the scow, five skiffs, and deck-loaded the tug with twenty- to forty-pound spring salmon—enough to run the cannery for three days.

"It was a pretty simple arrangement. They pursed with the steam anchor winch and gaffed the fish out of the net one at a time with hand-held cargo hooks."

A lot of good commercial fishers on the BC coast began their career in dugout canoes. Charlie Clarke was proud of such a beginning and the skills that he learned from his First Nations teachers. This canoe is fishing in the Somass River at Port Alberni in 1910, about the time Charlie was getting his first taste of salmon fever. (Leonard Frank/JHS 23497)

It may have been primitive, but it showed Peter Wallace and his brother what could be done with a purse seine net. The following year, 1912, they leased five seine boats from Washington State. The *W. No. 1* and the *W. No. 4* had open-base, 'Frisco Standard gas engines that leaked gas fumes into the crew's quarters in the fo'c'sle. The other three boats, the *Dexter*, the *Ferdelia* and the *Coho* had Atlas airtight engines that were a little safer. They had proper seine tables so that the net was set over a roller on the stern. Then the table turned so that the roller faced the side and the net was pulled back a little more easily. The net design was the same principle used today but they were smaller, only two 100-mesh strips deep and 200 fathoms long compared to modern nets for fishing inside waters which are 575 meshes deep and 220 fathoms long.

The boats fished so well that first year that the Wallaces bought all of them, complete with nets, for $3500 each just as the war was starting in Europe. Charlie enlisted in the navy and spent the war years down in Victoria, returning to Kildonan in December 1918. But instead of going back to the store, he went onto the *Coho* as engineer. His fellow crew members were natives, while the skipper was a Scot named Jock Sutherland. That winter they fished herring with the boat and the next summer they fished salmon.

It was at this time that Charlie encountered the Canadian practice of back-room resource management. "I was engineer for a year and then I went skipper. About that time they opened up certain areas that you could get a licence. Of course, all the canneries on the coast had certain areas nobody else could fish in. All the way right up to Rupert there were no private boats in the canneries. So they opened it up, and being I was in the navy I applied. From Bamfield to Sarrito was opened for a private seine boat. I had a Major from Port Alberni with me, but some Colonel back east got it. He didn't know anything about fishing. He got somebody to run it but didn't make any money because he didn't get a skipper that knew anything about it. If I'd got it I would have made a fortune out of it, because you sell your fish and nobody against you. Just one area all to yourself. But no way, I couldn't get in. So I went on fishing for Wallace."

CHARLIE CLARKE

If he couldn't be an independent owner-operator, he would be the best of the company men. Skippering the 52-foot *Ferdelia*, Charlie travelled the whole coast, learning the ways of the sea, the fish and the boats as he went. On one trip, he was traversing Hecate Strait, a dangerous crossing of several hours through exposed and shallow waters. "We were coming across from the Charlottes on her. All those boats burned gas or distillate and the gas tank was right in the bow alongside the galley stove. They vented right into the engine room. The gas tanks were so small that you had to carry drums of extra gas on deck. The engineer said that we were going to run out of gas. We hoisted one of the drums into the rigging and tied it down on both sides so that it wouldn't swing too much with the groundswell. Then we took the hose and put it into the bow tank. I was steering with the wheel that we used to have on the front of the cabin. The skylight was open when all of a sudden I heard a *poff poff*. The hose had come out and the gas had run on deck. The deck was leaking gas right on top of the stove and here we are in the middle of Hecate Strait with no lifeboat or nothin' in them days. It's a wonder that we didn't get lost or burned up in the early days.

"Another time we were coming across from Massett to Claxton, just the two of us, the engineer and I. It's just eighteen miles across from the spit but a southeaster was blowing. All you had was a back-breaking pump like you had on the gillnets. The engineer was pumping all the way until we got in behind Lucy Island. Then we got into Claxton and tied to the dock but it was blowing so hard in there that we banged up against the dock and took the guardrail off. It was the next year before we got up on the ways and the carpenter found the trouble was that the only thing holding the rotten wood of the stern together was the paint."

Navigation equipment, other than a compass, was nonexistent. "The compass was never adjusted," Charlie said. "The only way we could travel was to take our courses in the daytime, mark them down, get your tides and your wind and everything. Every day you mark-ed them down in the log because you couldn't take them off the chart or you would be way out. I came down to town [from Barkley Sound to Vancouver] once and the adjuster came on board and said that the compass was out two points."

Those first seine boats were good sea boats and the *Ferdelia* was fast for her time, but they had very small cabins compared even with the boats built in the 1920s. The deck cabin contained just the wheelhouse; the galley and crew bunks were down below in the fo'c'sle. There was a secondary steering wheel mounted on the front of the wheelhouse so that the boat could be steered from outside. This gave the skipper better visibility to watch for fish. Each species of salmon shows on the surface in

An early seine skipper steers the boat from the wheel mounted on the front of the wheelhouse. This allowed him to spot fish and to see the net being set.

By the late 1920s, when this photo was taken at Kildonan, the original US boats, including the *Dexter* on the left, had been joined by Canadian-built boats like the *W. No. 10,* centre and the *Talamaso,* right. The *Nahmint* appears to be rigged as a fish packer in this photo. The sunken scows in the foreground were filled with a mixture of copper sulphate and water for "blue stoning" the cotton nets to prevent rot.

Charlie's *Western Producer,* seen here rigged for drum seining at Quathiaski Cove on Quadra Island in the 1970s, became one of the best known boats on the coast. He fished her until his retirement and maintained an active interest in the industry from his Steveston apartment until his death in 1987. (VMM; AHB)

its own manner. Sockeye jump clear of the water and slide back in on their sides while humpback often just roll their dorsal fins in an action called finning. Coho jump clean, but they also fin. It is the skipper's responsibility to find the fish and Charlie learned early that this can be a lonely job.

"I remember once in my first year as skipper we were fishing in Quatsino. We didn't have much of a net, just three strips and 175 fathoms long. That's all you could carry on those little boats. We were fishing at Marble Creek for coho. They were finning and jumping all over the place, but we couldn't get a fish. So I started up the inlet. You stood on the outside of the pilothouse steering backwards with a wheel that came through the front of the pilothouse. The skylight to the fo'c'sle was right in front of me and I could hear my crew talking down below. I was the youngest one on the boat but they said, 'Where's the old bugger going to now! There's no bloody fish up there.'

"I never said anything. We got up the inlet about seven o'clock and set. We got about 2000 coho. Then it was dark. We made a couple of sets in the morning and loaded the boat with coho. We took them to the cannery and went back up and got a second boatload. Then I said, 'I hear you guys were wondering where the old bugger was going.' Then I told them about the skylight being open."

The first year is a crucial test for a new skipper. Since crews are paid a share of the proceeds of the catch, a skipper who can't find fish will have an even harder time finding crew. Within a few years of the Quatsino incident Charlie had such a reputation that he not only had more than enough crew willing to follow him, but when he changed fishing grounds half the fleet would be right behind.

Charlie pioneered a good many innovations in his career, working first with the Wallaces and then with Ritchie Nelson to build one of the largest fish companies (Nelson Brothers) in the province. He bought his first boat, the *Port Essington,* in 1930, and built and owned four more seine boats. Charlie put the same thought and care into the design and rigging of each one as the most devout fly-fishers put into the selection of their rod and gear. When he died in 1987, he left permanent gaps in the collective memory of the coastal fishing industry.

EARLY SEINE BOATS

The little fleet of seine boats that Charlie Clarke remembers coming up to Barkley Sound in 1912 were not the first seine boats in British Columbia. Just when the first purse-seine net was set in the province has been lost to memory, but it couldn't have been much before 1910. Up until that time gillnetting in the rivers and some drag seining from the beaches of the estuaries had been the preferred method of catching salmon in numbers large enough to support a cannery. Purse seining for salmon became popular just south of the Canada–US border in the San Juan Islands as a means of intercepting Fraser River-bound sockeye salmon in the first years of the twentieth century. The earliest of these purse seines were set from rowed skiffs, then pursed by hand-operated winches on small scows. The crews, mostly Norwegian and Croatian immigrants, camped ashore on the San Juan Islands.

The introduction of gas engines brought the setting and pursing operations together on the forerunner of the modern seine boat. In marked contrast to fishing vessels in much of the rest of the world, the cabin and the engine of a West Coast seine boat were set forward on the hull with the propeller turned by a long shaft running through the fish hold. This provided ample space for a platform, or seine table, on the broad stern. Even the earliest boats appear to have had sterns formed from squared timbers fastened one on top of the other almost in the manner of

One reason for the long life of many seine boats is that they spend a good part of the year tied to the dock during the off-season. The *Cape Mudge* (right), moored here at the foot of Gore Avenue on the Vancouver waterfront in 1928, was built two years earlier. She has her secondary controls on top of the pilothouse while the *Sea Angel* and the *Seamark No. 1* still feature a wheel on the front of the pilothouse. The *Cape Mudge* has two sacks of coal, a coil of manila anchor line and an anchor stowed on her foredeck. The *Seamark No. 1* has her brailer rigged but no net on the table. In winter the cotton seine nets were stripped into parts and hung in net lofts to dry. (VMM/World Ship Society)

In the first years of this century fishers, many of them originally from Dalmatia but called "Austrians" in recognition of the Austro-Hungarian Empire, came from south Puget Sound ports to seine in the San Juan Islands. Here in 1905 a mix of rowed and powered seine boats ride at anchor in front of a tent village where the crews camped for the season. Note the hand winches for pursing the net on the scows and the powered winches on the power boats, which also have cabins. (WMA)

Paul Martinolich with his mother Frances and father Richard continue the family's tradition in seining. Richard's dad built one of the first owner-operated seine boats in BC, the *B.C. Kid*, shown below (photo by Lloyd Fuerst) after being rebuilt in 1940. More recently she has been renamed *Homelite*. (AHB)

a log building. Shaped to extend out from the waterline, this massive stern structure was supported by a long horn-timber running up at an angle from the keel. This design extended the stern well out beyond the rudder and propeller to reduce the chance of fouling the net.

Point Roberts is the northernmost point at which Fraser-bound salmon can be intercepted by American fishermen. This was a favourite fish trapping location for US processors, and it is still an important seining location. The first purse seine set in Canadian waters probably was made by an American fisherman who couldn't resist the excitement of a school of sockeye in their distinctive sliding-jump display. With such temptation, more than one fisherman has been known to pay more attention to the fish than to the boundary.

The ports of Tacoma and nearby Gig Harbor have long been the Puget Sound base of Croatian fishermen and boatbuilders. Vananzio Martinolich came there from the Island of Losinj in the Adriatic Sea. He learned salmon fishing from his countrymen before he followed the salmon north to Point Roberts in 1891, crossed the Canada–US border off Tsawwassen, turned into the southernmost entrance to the Fraser River at Canoe Pass and settled just upriver at Ladner.

From his base in Ladner, Vananzio founded a family tradition of excellence and innovation in seine fishing which has continued to the fourth generation. In 1986, Richard Martinolich, who had just retired from his own fishing career, sat at the kitchen table of his Ladner home and recalled what he had heard of the earliest days of seining. "Grand-

father was in the towboat business to start with. Then they built scows and they had one scow for the family to live in and one for the net and they towed them across the gulf. That was in the year 1906 or 1907 and he had a licence then. The net was on the scow with a hand purse winch."

Richard went on to explain that his dad built two towboats for this scow seining, the *B.C. Boy* and the *Eva*. His dad fished them around Ladysmith and Nanaimo for dog salmon. But they weren't designed as seine boats. "I think it was somewhere around 1910 or 1911 that the *Yankee Boy* came up from the States. Our relatives had a big boatyard in Tacoma and another part of the family had a yard in San Francisco," Richard said, adding that the first seine boat his family actually built in BC was the *B.C. Kid*. She was built just downriver from Ladner at Port Guichon in 1913.

Richard's father also owned a boat called the *Amelia G* about this time, but Richard wasn't sure where it was built. The next seine boat his dad built was the 65-foot *Green Sea* in 1919. She sank in Barkley

EARLY SEINE BOATS

Sound with a load of pilchard shortly after the family sold her in 1924.

Purse seining had developed in Washington State to intercept the Canadian sockeye, but in Canada river concentrations made it possible to fill the cannery needs with sockeye gillnetted in the rivers. On some of the smaller up-coast sockeye runs, such as the Nimpkish, drag seine permits had been granted, often as a co-operative venture between an Indian band and a canning company. The drag seine was set from the shore. It didn't employ rings and purse line as it was set in shallow water and hauled from both ends back onto the beach with the fish trapped in the centre.

As pressure on the sockeye stocks increased, both through fishing and habitat depredation such as the Hell's Gate slide of 1913, the demand for less valuable species of salmon increased. Humpback or pink salmon were being canned along with coho and spring. The dog (chum) salmon which ran later in the fall was being salted for export. These were the species that the Wallace Brothers were fishing at their cannery in Barkley Sound. The Martinoliches were most likely fishing dog salmon on the east coast of Vancouver Island.

Up at Quathiaski Cove on Quadra Island, Reginald Pidcock, a former Indian agent, built a cannery in 1904 to take advantage of the labour pool provided by the Cape Mudge and Campbell River Indian communities. It was supplied with fish from the Campbell River, which did not have a sockeye run of any consequence. Fish also came from the troll fishery off Cape Mudge. Chief Harry Assu, who was born in 1905, recalls: "When I was twelve years old, right down on the beach here at low water slack, me and my older brother and my younger brother we went

At Quathiaski Cove W.E. Anderson, second from right in this 1913 photo, recognized the value of seine boats to intercept the sockeye salmon coming down through Johnstone Strait. He went to Atagi in Steveston to have a series of seiners built for the 1912 salmon season. (CRMA)

The *Quathiaski No. 8*, built by Atagi in 1916, was still seining salmon in Johnstone Strait in 1946, probably with Chief Billy Assu as skipper. By 1993 she had been renamed *Thunderbird Spirit* and fished bait herring out of Campbell River with skipper-owner Charlie Basset. (FT)

down and the fish were chasing the herring and they are right on the beach. We got over thirty of them, coho, we just clubbed them. You could just see the fish going by, just solid. We got ten cents apiece for them."

Pidcock sold his Quathiaski Cannery in 1906 to T.E. Atkins, who resold it in 1908 to the Quathiaski Packing Company. In 1912 the plant was again transferred, this time to the Quathiaski Canning Company with W.E. Anderson as principal shareholder. Anderson operated the firm until the cannery and its fleet were purchased by BC Packers Ltd. in 1938. "They used to have a big square-stern skiff," Harry Assu recalls, "and they used to drag seine at the spit [at the mouth of the Campbell River] and at the Cape Mudge village too. I don't know what year Anderson had his first seine boat. It was about 45 feet long. They used to call it *Slapjack*. The crew had to sleep in the stern. Only three guys sleep in the front and three sleep in the back. They had to turn the [seine] table before they could get down to their bunks. I think it wasn't built first as a seine boat. The *Quathiaski #1* was a packer that they used to collect the fish from the trollers. After that they built the *Quathiaski #4* and it was the first seine boat that they had built. Then they built the *#6, #7* and *#8*. They started building up pretty

EARLY SEINE BOATS

fast. The *Quathiaski #5* was a packer but the *Quathiaski #12* was a seiner."

The people at the Lekwiltok village of Cape Mudge became great commercial fishers and have owned many fine power boats over the years. Jimmy Hovell, a noted seine skipper from Cape Mudge who retired shortly before his death in the mid-1960s, explained how the decision to adopt modern technology was made. When he was a young man, he, Johnny Dick and Harry Assu's father, Chief Billy Assu, made a trip in the early years of the century, paddling a canoe home from work on the Fraser River. They stopped for the night just north of the present-day community of Powell River, at the Salish village of Sli-ammon. Unlike their home village, this was a missionary village with a sawmill and frame houses in place of the split-cedar longhouses. The men were invited to stay overnight in one of the houses, and they were amazed to see a piano and other powerful symbols of the new white culture in the house. The next day on the long pull across the Strait of Georgia and north toward Quadra Island, they talked about what they had seen. The three young men were eager to try new things, but at the same time they were concerned about the diseases and alco-

In this 1912 photo one of the Atagi-built Quathiaski seiners brails sockeye in Deepwater Bay just north of Seymour Narrows. At this time First Nations people were not allowed to seine salmon so the crew is all Caucasian except for the Chinese cook. According to Charlie Clarke it was the Chinese cook's job when brailing to stand at the side of the cabin and pull the vang line to swing the boom. Because this was difficult to pull with the vang mounted on the gunwale, as it appears to be here, a cross boom was later added just above the cabin roof for the base block of the vang. This is still called the China boom. (CRMA)

hol brought to their land by the Europeans. When they landed at Cape Mudge, it was with a joint determination to grasp the white man's technology and use it to build a place for their people in the twentieth century.

One of the most attractive pieces of technology for men who had just pulled a canoe over 100 miles was the gasoline-powered boat. In 1985 Harry Assu recalled his father's purchase of his first one. "I would say I was about eight or nine years old when my father got his first gas boat. He was the first one to get a gas boat, a 6-hp Fairbanks one cylinder. It used to be really noisy. It must have been about a 30-foot boat. He used that for gill-netting and trolling. He had it for two or three years and he got another one, a bigger boat."

At a number of locations on the coast Indian people worked with cannery owners to operate drag seines at river mouths. But Indians were not allowed to skipper seine boats. The seine boats that W.E. Anderson had built for his Quathiaski Cove cannery were skippered by Scots. Harry Assu explains: "You know us Indians we wasn't allowed to seine. I went with my father to the Cove, they had a public meeting there. This MLA for Alberni named Neil was there. My father asked if they could help us get a start seining. We were only allowed trolling and gillnet. This guy said, 'I'm going to Ottawa. I'll take it up.' Two or three weeks' time my father had a letter from him saying, 'You can go ahead and start seining.'"

Within a decade of buying their first power boat, the people of Cape Mudge were using the most advanced fishing technology of the day.

"In 1922 all the Indians started seining. My father was running the

Quathiaski #14. I fished with Jimmy McPherson, a Scotchman, he had the *Quathiaski #12.* I fished with him for four years and then I went to fish with Jimmy Hovell for one year. We had the *Annandale.* Jimmy Hovell started in the fall and we fished chum salmon. There were a lot of fish in Alert Bay in 1924. We were there three weeks and we put in over 50,000. I guess the lowest day was 5000. We were fishing for the Howe Sound Fisheries. We had three packers following us all the time. We just brailed right onto them and every day we filled them up. The Howe Sound Fisheries were salting it in Blind Channel and shipping it to Japan. The company was owned by brothers, Fukuyama and Sugiyama. They had six boats all named *Howe Sound* and they chartered the *Annandale* from Bell-Irving. We fished the chums up all the inlets and even in the straits. I remember once on the 15th of November in Deepwater Bay we made a set in the morning and got 2000. In 1926 Jimmy Hovell had the *Sea Luck,* a brand new boat, and we went to Loughborough Inlet to fish for pink salmon. We made two sets and we got 45,000. This Canadian Fish boat was there, George Skinner, he made three sets for 90,000."

Seine boats were introduced to BC's West Coast about 1912. By the 1920s seiners were being used widely for the salmon fishery and for the herring and pilchard fisheries as well. Introduced first in the Gulf of Georgia around Nanaimo, they made their way to the outer coast in Barkley Sound and north into Johnstone Strait. Their advantages in catching the lower value species of salmon were immediately apparent and they quickly became important in the valuable sockeye fishery. With the advent of seine boats, the technology was now in place to exploit, and over-exploit, the fishery resource that had sustained the First Nations people for thousands of years.

The Canadian Fishing Company had the *Tasoo* built in 1917. She had a large house for the day. The engine was still exhausted out the side of the hull. On later boats a funnel was added to enclose a muffler and exhaust the engine through the cabin top. (VCA)

ATAGI, BOATBUILDER

Tsunematsu Atagi arrived in Steveston from Japan in 1900. This was in the era when Fraser River canning companies routinely played off the three major ethnic groups—native Indian, European and Japanese—against each other in order to prevent the organized negotiation of fish prices. As a result, prices remained very low. Since salmon fishing was seasonal work anyway, a man who wanted to make a future for his family in this new land required something more. For Tsunematsu, this was boatbuilding.

In 1905 he opened his own yard. By that time waterfront property was already at a premium. Some time before the Atagi yard was set up, the Scottish Canadian Cannery was built west of Garry Point on the grass-covered tidal flats. To reach it, a bridge or causeway had been built from the point where the Steveston dyke turns north. It was also here that the south arm training jetty had been built out from the dyke to arch westward toward the light ship. In the space between the causeway and the jetty Mr. Atagi built his boatyard.

His first boats would have been the Columbia River-type gillnetters of the day. The Japanese technique of building these was to lay the keel, set the molds, then tack the planks to the molds. Sometimes a saw cut was made between the planks with a Japanese-style saw that cut on the push rather than the pull stroke. This cut ensured a perfect fit which required

Inside the Atagi Yard at Scotch Pond in Steveston one boat is nearing completion while the horn-timber has been mounted to the deadwoods of what will be the stern of a new boat. (Atagi family)

Ray Hamata took this photo in his father's boat shop at BC Packers' camp on Westham Island about 1926. It shows a gillnetter under construction. The boats were built by setting up hull-molds every six feet and then setting the cedar planking onto them. With the planking in place the oak ribs were steamed in, using one continuous piece right around the hull. This made for rapid construction and tight-fitting planks. At the time a 30-foot gillnetter sold for $450 with the purchaser supplying the engine for the builder to install. (R. Hamata)

only a string laid between the plank edges for caulking. The ribs were bent into the hull after the planking was in place. In this way a single rib could be bent right around from port shear plank to starboard shear plank.

The Atagi yard prospered and was soon building larger packers with fine, upswept sterns. In 1912 the growing interest in the new seine technology gained the Atagi yard the first of a series of orders for several seine boats from the Anderson Cannery at Quathiaski Cove. These were well known on the coast for years as the *Quathiaski* boats. Probably the only one of the series not built by Atagi was the American-built *Quathiaski #12*. It is tempting to speculate that an Atagi-built boat was one of the first Quathiaski seiners. Veteran Cape Mudge fisherman Harry Assu recalled that the first group of seiners included a packer with a seine table built onto the stern. However, the crew's quarters were in the stern, and the table had to be turned aside every

ATAGI, BOATBUILDER

Launch day at the Atagi Yard shortly after 1900. The yard was located at the southwest corner of Lulu Island and the water in front of it was very shallow. The boats were launched bow first so that they could be pulled out through the delta mud. The walkway in the foreground went out to the Scottish Canadian Cannery built on piles at Garry Point. (Atagi family)

night so that the crew could go below.

Because of the mud flats and the lack of dredging, it was always necessary to launch the boats bow first at the Atagi yard so that they could be dragged through the mud before floating free. The causeway, which ran several hundred yards out to the cannery, served as a convenient point from which to pull the reluctant craft into the water.

The list of boats built at the Atagi yard is very long; many are well known on the coast. *J.I.4*, owned for many years by the Moses family at Alert Bay, was built in 1912 for the Jervis Inlet cannery and served first as a packer and gillnet tow-out boat. Others include the *Arieta* (1916), *Kiona* (1925), *Chief Y* (1926), *Shuchona* (1927), *Maple Leaf C* (1929) and *Mermaid* (1938).

Over the years Tsunematsu's boatyard grew just as his family grew. He built a house on the dyke near the boat shed, and the cherry trees he planted along the dyke still stood in 1992. His three sons—Kaoru, Hisao and Kenji—worked in the yard and fished the Fraser and Rivers Inlet. Hisao, who retired from fishing in 1990, remembers those long trips north in the summer when they always stopped to visit an uncle who had moved to Quathiaski to work for Anderson in the boatyard there. He also recalls that Scotch Pond was much more exposed than it is now. The

The Nakade and Tabata Boatyards in Steveston about 1918. Uchi Nakade is at the top left and his partner Mr. Tabata at the top right. Tabata later went on to build herring salteries in the Gulf Islands and Uchi Nakade built seiners for him. The mound of earth in the foreground is the dike which separated the buildings from the Fraser River. The two large vessels under construction with the sawn frames and very fine sterns appear more like yachts than fish boats. Two Columbia River-type gillnet skiffs are in front of the two big boats. (Nakade family)

whole of what is today Garry Point Park was low, grass-covered mud flats. When a westerly blew, the driftwood pushed right up to the yard. It was these drift logs that his family pulled up on the beach and on which they built their house.

Altogether they had about three acres of land. They had lived there, and built a fair percentage of the BC fishing fleet, for more than thirty-five years when war broke out between Japan and Canada. The politicians of the day, with encouragement from some elements in the fishing industry, took the opportunity to seize the property of native-born as well as immigrant Japanese fishermen. The Atagi family lost everything. They moved to Celista near Salmon Arm, where they stayed until they were allowed back on the coast.

Tsunematsu Atagi, pioneer Canadian boat builder, died in 1949 shortly after his family's return to the coast. The sons took up boatbuilding again, at a different location in Steveston, and did much to rebuild the BC fishing fleet through the 1950s and 1960s.

FROM THE ADRIATIC

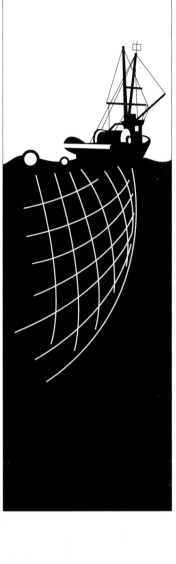

T he sea can be a dangerous place and fishing is hard, dangerous work that discourages experimentation and exacts harsh charges for errors. Fishing communities of the world tend to the conservative: traditions serve well and technological change must prove itself at someone else's expense before the individual fisher will risk his life and livelihood to accept it. But the seine fishery of the BC coast was an exception to this rule. Aside from the First Nations people, everyone came from somewhere else. Having already made a commitment to change, they were more willing to adopt new techniques.

The first wave of newcomers from the coast of Dalmatia on the east side of the Adriatic settled around Tacoma and Gig Harbor, then came north to fish salmon in the San Juan Islands and across the border in Canada. A second and larger migration of Croatians came directly to Canada in the 1920s. Dick Anzulovich and Louie Percich, who belonged to this second group, came from fishing families on the same small island in Dalmatia, but it was in Vancouver that they met and formed a friendship that carried them through a half century of Pacific Coast seining.

Louie left home at fourteen to work as a crewman, first on a yacht and later on Italian and Swedish freighters. Seeing the world from the deck of a freighter, Louie realized that much of it looked more promising than his Dalmatian home. The first time he jumped ship was in Chile where the islands strung along the coast reminded him of home. But it wasn't right

for him, and he hired onto a Swedish ship that took him back to Europe. The next time he jumped ship was in Prince Rupert. "I had no money, not one single cent. I couldn't speak one word of English." He ended up hundreds of miles inland at Prince George washing dishes in a hotel for room and board. He spent over a year there, but he was a seaman and a fisher so he eventually found his way to Vancouver.

Dick Anzulovich, a few years younger than Louie, had already caught the fishing fever from his father before leaving Dalmatia. They went out at night into the Adriatic to spear fish attracted to a light. He saw a few small gillnets and

Louie Percich in 1987. (AHB)
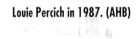

drag seines there but no purse seines, nor were there processing plants to take large catches. The Anzulovich family speared rock cod, squid, striped bass and other species, and sold them on the fresh market. Fish was their main source of cash. On a small piece of land the family kept three or four goats, a pig and a donkey. They grew almonds, figs and olives for their own use, and made wine for some cash sale. By the time the tax men were done with the family there simply was not enough left over for education or for the families of grown sons. Dick's father borrowed some money at 30 percent interest and sent his son to Canada.

At that time only a Canadian citizen could get a commercial fishing licence in Canada. Citizenship required five years' residence, a long time to young men who could neither read nor write and knew no trade but fishing. Dick arrived in Vancouver in June 1929 and moved in with his brother, who had arrived three years earlier and lived in a squatter's shack on Deadman's Island in Coal Harbour. Every morning Dick rowed a skiff across the harbour and walked the waterfront in worn-out shoes with cardboard insoles. There was just enough work to stay alive and off relief, but there were times when he would have returned to Dalmatia if he could have afforded a ticket. His salvation was a salmon seine boat looking for a crew, and the secondhand licences for sale in Skid Road beer parlours for five dollars.

"I went with my Uncle on the *Cape Henry* for Canadian Fish Company," Dick remembers. "Somebody reported us, me and Louie Percich, that we were fishing with no licence. Old Cameron was Fisheries patrol man. He came on the boat and said to bring the licence. He look at it with his glasses and then looked at me over his glasses. My licence was for an Irishman and the officer, he asked me, 'What part of Ireland are you from?'

"I told him, 'Dalmatia.'

"Seven days in jail or twenty-dollar fine, so my uncle paid the fine and kicked us off the boat. We had to go home."

Louie Percich recalled the same incident. "I had bought my licence for five dollars from some fellow on Skid Road, it was in the name of Dickson. The Fisheries officer came on the boat and said, 'Is your name Dickson? How come you speak Yugoslav, and you don't speak English?'

"I said, 'Mr. Cameron, my father was a sailor. He come over to Dalmatia, met my mother, marry her and took her to England. Then I was born in England, but my father died so my mother went back to the Adriatic.'

"He looked at me for a moment, then he said, 'Your story may be right but you don't look like an Englishman.' So he took us into Butedale. They fined me and sent me back to Vancouver but I got on another boat."

Louie also went fishing with an Italian named Frank Mangarella. He worked on Mangarella's two boats: the *Adele M*, named for the owner's wife, and the *Frank AM*. Louie worked on a number of boats and before the five years that qualified him for citizenship were up, he was skipper of the *Z Brothers*. "When the Fisheries officer come, I run in the engine room and hide and Zovich, the owner, was the skipper."

Dick Anzulovich took whatever fishing work he could find, but always seining. "I fished smelts at Port Moody up Burrard Inlet in an old

Louie Percich skippered the *Z-Brothers* before he had his Canadian citizenship. The boat was built by Tara Brothers in Ladner in 1929. (VMM)

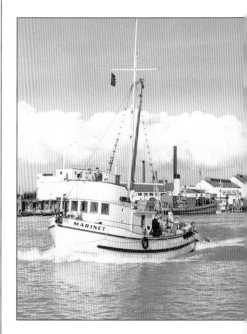

gillnet boat. We had a little purse seine that we pursed by hand. We sold the fish to Chinese for fifteen cents a pound and any that were left over we sold to the Canadian Fishing Company at the foot of Gore Avenue for ten cents. In 1930 I went out on the *Neptune No. 1* with my uncle. It was a brand new boat. In 1936, 1937 and 1938 I skippered the *Menzies Bay* for the Canadian Fishing Company. In 1939 I bought the *Faith of Sechart* from Nelson Brothers Fisheries. It used to be a two-masted boat but they took one mast off. In 1940 Ritchie Nelson chartered my boat for herring but he said to me, 'Dick, on the herring you got to take all skippers, no crew.' Mate Bobic, who got the *Western Breeze* after, he was with me and Louie Percich, George Brajcich, Johnny Anzulovich, John Selic from Ladner, and Bill Pitre was skipper and we were all crew. Then in 1941 I skippered the boat on herring. I took some skippers but not all of them. I told Ritchie that I wanted some of my own crew or I wouldn't have my crew for the salmon the next year. I kept that boat until 1946 when I bought the *Rainbow Queen* [ex *Howe Sound*]."

There were relatively few seine boats employed in the winter herring reduction fishery. For this reason a fishing company like Nelson Brothers would insist that a herring crew be composed largely of the company's salmon skippers. In this way they kept their good people employed and avoided raids by competing companies on the ranks of their skippers.

Louie Percich started early with Nelson Brothers also and skippered a

George Brajcich was a contemporary of Dick Anzulovich and Louie Percich. The Croatian community has made, and continues to make, major contributions to the BC fishing industry. (FT)

Both the *Sea Master*, built at Sterling Shipyards for Frank Tomasich, and the *Nanceda* were fishing their first season of herring when this photo of them working together was taken in 1950. (JW 1493)

number of boats for them, including the *Ribac, Western Commander*, and *B.C. Pride*. He bought the *Sea Biscuit* (ex-*Izumi VI*) from a Japanese Canadian fisher about 1935. "I kept it for some years," Louie said. "Then I sold it to the company. In those days the fish companies were different from now. They served you. Of course they helped themselves to help the industry, to build the industry. If you were young and ambitious there was no problem about money. Ritchie Nelson was a very good man. When I built the *Nanceda* in 1951 I had $100,000, no interest. It was different than now."

With the *Sea Biscuit*, Louie fished only salmon. Her registered length of 51 feet was big enough when Louie's old boss, Jisaburo Kasho, had used her to fish herring as a double seiner with the fish all going into scows. By the time the Euro-Canadian fishers were fishing herring in the late 1930s and 1940s, they were using small seine boats like the *Sea Biscuit* only as tow-off boats to assist the larger seine boat when its net was out.

"Dick Anzulovich and I fished all over the coast. We fished the Queen Charlotte Islands before there were charts. We used to go into an inlet and Dick would say, 'You go first.' I would say, 'No, you go first.' There were no echo sounders then, just a line with a piece of lead to measure the depth."

It took courage for these men, and others like them, to leave their Mediterranean homes for distant Pacific shores. That same courage and determination drove them to excel in the Canadian fishery. In the modern fishery, many of the children and the grandchildren of these pioneers continue to lead. Others have used summer fishing dollars for training in the professions and business. But no matter how successful the second generation may be, it is the courage of the ones who travelled here to develop a new fishery in a new land that is remembered when stories are told around the galley tables of the West Coast.

PADDLE WHEELS ON THE PACIFIC

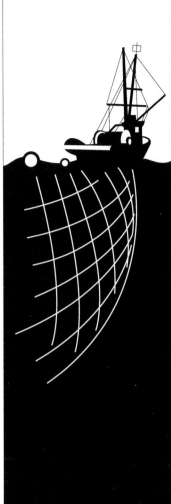

In the outports of Norway's west coast the old cod-drying racks stretch in rows on expanses of flat land among the grey rocks. They are seldom used today, but the sound of dried cod carcasses blowing together in the wind still seems to reverberate in musical tribute to the fishermen who have worked the North Atlantic waters for thousands of years. It is a hard land bordering a harsh sea, a place where incredible feats of labour only increased the chances of survival. In the cold northern air, even the hay that is harvested on fields sloping down to the sea must, like the cod, be hung to dry. The sea and fishing are bred deep in the coastal Norwegians.

When Edgar Arnet's dad left this hard coast for a new life in America in about 1892, he landed in New York. He and his partner got work as carpenters in St. Louis. Recalling the history almost a century later, Edgar said, "In their younger days they had been fishing in the old country of course, so they didn't like being way inland. They wanted to get out to the coast." When they heard about Seattle, the young men followed the call of the sea out to the West Coast where they worked briefly as carpenters rebuilding Seattle after the great fire. "They heard about fishing out on the west coast of Vancouver Island. They went to Victoria and took a boat up the coast. Dad took up a preemption of 130 acres right where the airfield is back of Long Beach. They built a home there and they built their own fishing boats. They heard about the Fraser River being the big run of salmon during the summer months, so they rigged up some sails and oars and a couple of them together went down. There were only one or two lights on the coast then. They gillnetted on the Fraser, there were all kinds of fish then.

"In 1894 my mother came out from Norway and they got married in Victoria. After that every summer my father and his partner would sail, or row if there was no wind, down to the Fraser River. In about 1900 an outfit call Clayoquot Sound Canning Company opened up a cannery right in Tofino, so they fished for them in the summer and fall for a few years. Then they got into building lighthouses at Pachena, Cape Beale and Triangle Island. That got them into about 1912."

The Clayoquot Cannery had been established to take advantage of the good runs of sockeye, coho and chum salmon into Kennedy Lake near Tofino. In the early years the cannery was supplied by drag seines worked from the beach and from gillnet boats towed out into the inlet by a tender that also carried the salmon back to the plant. By 1913, when he was thirteen years old, Edgar was working alongside his dad on some of the

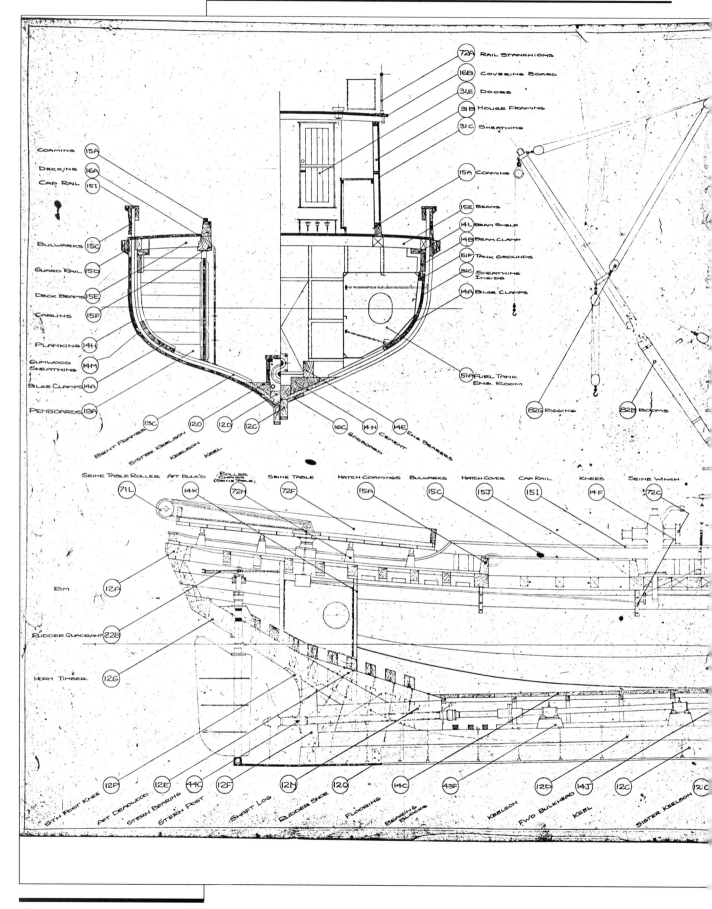

Coming (15A)
Decking (16A)
Cap Rail (15I)
Bulwarks (15C)
Guard Rail (15D)
Deck Beams (15E)
Carlins (15F)
Planking (14H)
Gumwood Sheathing (14M)
Bilge Clamps (14A)
Penboards (19A)

(72A) Rail Stanchions
(16B) Covering Board
(31E) Doors
(31B) House Framing
(31C) Sheathing
(15A) Coming
(15E) Beams
(14L) Beam Shelf
(14B) Beam Clamp
(51F) Tank Grounds
(31C) Sheathing Inside
(14A) Blge Clamps
(51A) Fuel Tank Eng. Room
(82G) Rigging
(82B) Booms

(13C) Bent Frames
(12O) Sister Keelson
(12D) Keelson
(12C) Keel
(16C) Garboard
(14N) Cement
(14E) Eng. Bearers

Seine Table Roller (71L)
Aft Bulk'd (14K)
Roller Chafers (Seine Table) (72H)
Seine Table (72F)
Hatch Coamings (15A)
Bulwarks (15C)
Hatch Cover (15J)
Cap Rail (15I)
Knees (14F)
Seine Winch (72C)

Rim (12A)
Rudder Quadrant (22E)
Horn Timber (12G)

(12P) 5th Post Knee
(12E) Aft Deadwood
(14C) Stern Bearing
(12F) Stern Post
(12H) Shaft Log
(12Q) Rudder Shoe
(14C) Flooring
(43F) Bearing Block
(12D) Keelson
(14J) Fwd Bulkhead
(12C) Keel
(12C) Sister Keelson

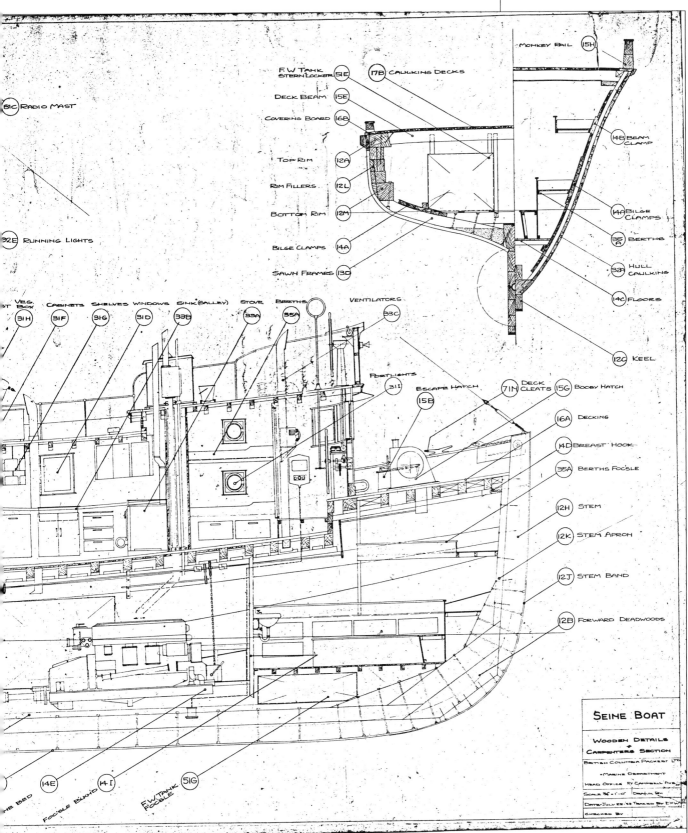

BIC Radio Mast

32E Running Lights

VEG. BOX
ST BOX 3IH

3IF Cabinets

3IG Shelves

3ID Windows

33B Sink (Galley)

33A Stove

35A Berths

33C Ventilators

3II Portlights

FW Tank Stern Locker 5IE

Deck Beam 15E

Covering Board 16B

Top Rim 12A

Rim Fillers 12L

Bottom Rim 12M

Bilge Clamps 14A

Sawn Frames 13D

17B Caulking Decks

15H Monkey Rail

14E Beam Clamp

14A Bilge Clamps

35 Berths

33 Hull Caulking

14C Floors

12C Keel

Escape Hatch 15B

7IN Deck Cleats

15G Booby Hatch

16A Decking

14D Breast Hook

35A Berths Fo'c'sle

12H Stem

12K Stem Apron

12J Stem Band

12B Forward Deadwoods

14E

14I

F.W. Tank Fo'c'sle 5IG

Fo'c'sle Bunks

BE Bed

SEINE BOAT

WOODEN DETAILS
&
CARPENTERS SECTION

BRITISH COLUMBIA PACKERS LTD
—MARINE DEPARTMENT

HEAD OFFICE FT CAMPBELL AVE

SCALE ⅜" = 1'·0"   DRAWN BY

DATE JULY 22 '53 TRACED BY E.F.W.

CHECKED BY

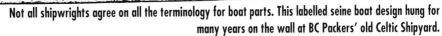

Not all shipwrights agree on all the terminology for boat parts. This labelled seine boat design hung for many years on the wall at BC Packers' old Celtic Shipyard.

first seine boats used on the coast. "They were open seine boats, built right there at the cannery. The boat that I was fishing on was called the *Bison*, and we had two other boats. The engine was in the forward part of the boat and the shaft went back to the propeller. In the middle of the boat they had a clutch so that the winch worked without the propeller turning so that you could purse the net. The winch was chain drives with a couple of heads on a 2-inch shaft in a wooden frame. They had a cabin over the engine with a little deck forward. The middle part was all open and on the stern was a deck and turntable that we piled the net on."

The cannery also had a tender for packing salmon from the drag seines to the plant. It was on this boat that young Edgar went halibut fishing, first off the West Coast, then north into Hecate Strait and Alaskan waters. Here was a fishery to match his dad's tales from the North Atlantic. "My uncle helped install the engine so the first couple of years the owner used it for packing, but then he decided to take her out halibut fishing and my uncle helped me get on as deckhand and second engineer. The fishermen in those days used to get sores in their hands from hooks.

The *Cape Beale* has the classic hull form of a halibut boat. The long fine bow will break mountainous waves, and the fine rounded stern will handle following seas.

They couldn't go out in the dory so the skipper got me to take the place of anyone that was hurt. That was the first halibut fishing I ever did. It was 1916 and I was sixteen years old."

By 1924 Edgar had decided to build his own boat. With a partner, John Berg, he had a Norwegian immigrant draw up a set of plans and make a model for a halibut boat. Because they must often work in heavy seas to haul the longlines to which the short gangen (lines) with their hooks are attached, the sea-keeping qualities of the boat are all important. After some debate with the builder over the best design for the stern, they went ahead with the construction of a classic halibut boat. A small cabin was set well back from the bow, but in a departure from tradition the cabin was ahead of the hatch. The long propeller shaft was in two parts; the main shaft which ran through a series of bearings under the floor of the fish hold was coupled to the tail shaft which extended out through a stuffing box and the heavy timbers that formed the keel and stern post. A steel shoe ran out from the end of the wooden keel to hold the bottom of the rudder in place.

The boat was built in a rented Vancouver shipyard near what is now the north end of the Burrard Street Bridge. By the end of February 1925, they had the hull planked, caulked and painted. The deck beams were in place, although the deck hadn't yet been laid. After launching, they towed her around to Coal Harbour and rented space at Andy Linton's boatyard where they completed the decking, built the house and rigged the boat in time to leave for the Alaskan halibut grounds on the 3rd of June. They picked up crew, ice and bait in Prince Rupert and headed out to fish around Kodiak Island, Portlock Bank and Yakutat. John Berg, who had made a couple of trips to these waters on a small Seattle boat, served as skipper.

The following summer, in what was to become a familiar pattern, they got work for the boat packing salmon before going out to fish halibut again. Halibut could be fished virtually year-round with a ten-day layover between trips. John Berg died in 1927, so it was Edgar running the boat on that fateful trip in 1928.

They had been fishing out around Kodiak Island. Fishing had been good and when time came to deliver the trip, Edgar set a course directly across the Gulf of Alaska for Dixon Entrance at the top of the Queen Charlotte Islands. "We were just about in the middle of the Gulf of Alaska, over 250 miles from the nearest land. The darn thing broke right in the shaft log where you couldn't get hold of it. The propeller slid back from the speed that we were making and jammed up against the rudder. The rudder was useless. So then I thought of this side-wheeler. I talked it over with the crew and they thought I had something wrong upstairs.

"We were seven men, some of the crew were in favour of trying to work out something. So we started to make this side-wheeler. Three of the crew wanted to take off in the dory and row for land. We objected to that, so they couldn't. We worked on this jury-rig. It took us about two and a half days until we finally got it going.

"By golly, we were making about two miles an hour. This winch we had, you could turn it so that the heads were fore and aft or athwartships. This made it a lot easier. We used a rope messenger from the winch head to a drum in the mid-

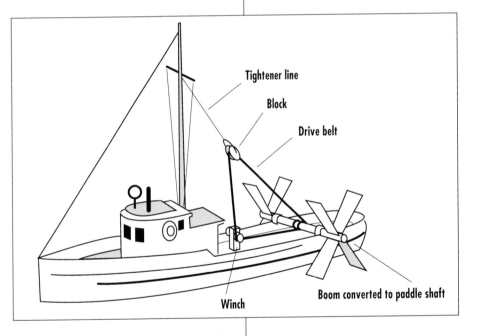

Tightener line

Block

Drive belt

Boom converted to paddle shaft

Winch

dle of the boom which we had laid across the gunwale to use for the shaft with paddle wheels on each end. We took a couple of turns of rope around the drum and the winch with an endless rope. Well, the darn ropes wouldn't last forever. A couple of hours and they were all chaffed up. It was our tie-up lines that we were using.

"When they were all gone we took the cable off the anchor winch. It was quite flexible, not too stiff, so we rigged that up with a couple of turns around the drum and the same on the winch with a tightener up the mast for the slack. This was just a snatch block on the end of a line for the cable to run through. The line ran through another block at top of the mast and could be tightened from the deck.

"It worked fine, but then it started to chew up the wood on the boom that we were using for the paddle shaft. We saw that wouldn't last very long so we thought of the funnel that the exhaust pipe went through. It was about two feet in diameter. It wasn't welded, it was rivetted, so it wasn't hard to knock the rivets out and wrap the sheet metal on the drum. That went fine, but the steel cable would wear the eighth-inch metal, so we would move it over. In a couple of days there was no more funnel left, it was all chewed up.

"We had a bunch of fishing anchors with one-inch stocks. We took these and lashed them about four or five inches apart, like a Jacob's ladder. We wrapped that around the drum and it worked fine. You could go a couple of days before moving the anchor stops.

"We had tried to make course for Dixon Entrance but the darn wind was against us and was pushing us up into the Gulf so we decided to make for Sitka. We still had a southeaster and one day it was blowing about thirty or thirty-five miles an hour. It got quite rough and with the boat rolling the paddles would dig way down in the water. On one roll that she took the boom that we had made for the paddles broke off on some cross grain. So we had to put our dory overboard and fetch it back. We managed to bolt it and tie it together. The paddles were too big so we cut them down to half the size.

"The winds were still taking us up the Gulf and we couldn't make Sitka. On the seventh or eighth day we saw the top of some mountains so then we knew we were getting close to the east side of the Gulf. In the evening we saw a boat in the distance. The tide and wind were taking us right for him. We got there before dark, launched our dory and rowed over to him. It was an American boat from Petersburg called the *Baltic*. They'd had a good day's fishing and had gone to bed. The skipper got up and I told him what went wrong. 'Well,' he said, 'there's a passenger boat coming down the coast tomorrow.'

Built in 1959 and named for one of the Aleutian Islands, the *Attu* fished halibut in the Bering Sea. She is now limited to Canadian waters by the 200-mile fishing limits introduced in the 1970s. She seines salmon and herring and moors at the False Creek Fisherman's terminal in the off-season. As with so much of the working waterfront, apartment buildings crowd the shores, making moorage for fish boats scarce. (AHB)

"But I didn't figure he'd stop for us so I asked the skipper of the *Baltic* if he would tow us to the first cannery down the coast. This was the McNeil & Libby cannery about a hundred miles south and just in from Cape Spencer. He agreed so we dismantled the paddle wheels and he put a line on us. I thought we would get one of the cannery tenders to tow us up to Juneau, which is only about seventy miles. But the cannery manager said no way he was going to let any of his tenders go as he was too busy getting his traps set. So I talked the skipper of the *Baltic* into towing us up. He was reluctant because he was on good fishing. But I told him the insurance would pay.

"In Juneau we sold what fish we had—we had a fair trip—and contacted the insurance. They allowed us to have a new shaft installed and paid for the tow. They didn't pay for a new boom or funnel. The newspaper people in Juneau got wind of our trip and wanted us to rig the paddle wheels up again so they could take pictures of it. But we had taken it all apart. After a week in the shipyard we were ready to head back out to the Kodiak grounds.

"On the trip when we broke the shaft we were fishing alongside an American boat called the *Grant*. The skipper of that boat always sold his trip in Rupert. When we left the fishing grounds and set course for Prince Rupert we figured he'd be leaving the same day, but we never saw him after we broke the shaft. He could have passed just a few miles away and we wouldn't have seen him.

"Twenty-five years later, about 1952 or 1953, the same skipper on the same boat was fishing off Kodiak Island. He was setting a course for Dixon Entrance. When he was halfway across, just about where we snapped our tail shaft, he came on the air with a Mayday call. His crankshaft had broken. The US Coast Guard came on the air right away. The skipper of the *Grant* gave his loran position and within twenty-four hours the Coast Guard had come, put a line on him and towed him into Sitka or Ketchikan. We didn't even have phones in the 1920s. It's a lot different today."

Edgar went on to have a long and successful career in the fishing industry. He built the combination seine boat–longliner *Attu* with a steel hull and aluminum cabin in 1959. One of the first steel seiners built with an aluminum house, she is still fished by Edgar's son, George. But when halibut stories are told in the galley, one of the favourites will always be the tale of Edgar's paddle wheels.

HERRING SALTERIES ON THE GULF ISLANDS

"That was the worst occupation, that and the shingle mill," says Ted Nakatsu, recalling the three seasons he fished herring for the Gulf Island salteries. Maybe that is why so few people know about this fishery today. Hundreds of people were employed catching and salting the herring for export to China. Some fortunes were made, and a fine fleet of boats was built. A good many of the boats are still fishing today, but the fortunes all were lost when Japanese Canadians had their property seized prior to their internment in 1941 and 1942.

The herring fishery started before World War One with rowed seine boats in Departure Bay, near Nanaimo. These rowed boats, like the later, powered boats, fished in pairs with a large purse seine net piled half on each boat. The boats encircled the herring from either side and set the net between them.

By the 1920s, the Steveston boatyards of builders like Atagi, Nakade and Yamanaka were busy building boats for the several salteries

In the years before World War One Japanese immigrants began exporting salted herring to China. The fish were caught from rowed boats adapted from Columbia River-type gillnet boats. In these photos taken in Departure Bay, some boats are tied together in pairs while small spotter boats look for herring flipping on the surface. In the foreground a pair of boats have begun to set their nets. (UBC)

that dominated the industry. Over on the north shore of Burrard Inlet at the mouth of Seymour Creek, J. Ashima was building the *Hatta* boats. Some of the salteries—Tabata's on Jessie Island in Departure Bay, for example—had their own builders, like the legendary Jirokichi Arimoto, who Ted Nakatsu recalls hearing "never measured anything. He could just go by eye."

It was widely thought that white men could not catch herring in the Gulf. That may be why, in the 1920s when regulations began to curtail the participation of Japanese Canadians in the Fraser gillnet fishery, and to bar their access to salmon seine licences, the same regulations ruled only that half the crews, and not the skippers, on the herring seiners had to be non-Japanese. Randy Thompson of Gabriola Island recalls that this happened in 1924. "We worked in Departure Bay the first year they brought that law in—my dad, my mother and us three boys. There was Tabata on Jessie Island. Then on Newcastle Island there was Tanaka, and next door was Kasho Camp, and then there was the big camp, Ode Camp, but I think Matsuyama was the name of the fellow that owned it at that time. Then there was Yip Sang with a camp next to it. Those were the five on Departure Bay. They were using twin seiners to fish. Tabata had a fleet of *Jessie Island* boats."

Shortly after Randy started fishing at Departure Bay, the stocks declined and the fishery there was closed. Additional salteries were built on various Gulf Islands. Their ownership is sometimes hard to trace,

Rikimatsu Tabata's herring saltery was on Jessie Island in Departure Bay. His daughter, Hideko, remembers watching from the family's house on the top of the island as her father rowed around in the bay looking for herring that his seine boats could set on. Today the island lies deserted under the gaze of passengers arriving on the ferry from Horseshoe Bay. (Hamagami family)

Yasaburo Hamagami gave his birth name Ode to his herring salteries. (Hamagami family)

In the 1920s and 1930s Canadian law defined the racial makeup of the herring crews. In this photo Caucasian-Canadian fishers brail herring into pot scows from a seine under the direction of Japanese Canadians. The bluffs in the background appear to be one of the Gulf Islands. (UBC)

probably because several saltery owners co-operated among themselves in ownership and possibly marketing. Randy Thompson fished at a camp on Reid Island, where Tanaka and Kasho also had camps. "Matsuyama, Yip, Tabata and one more were over on Galiano. When they closed the camps up at Departure Bay, Matsuyama turned his into a boatyard. He later had around forty seine boats. He used to lease them out to white skippers for salmon; then, when the herring started, he would put them on towing herring scows. We used eighty- or ninety-ton pot scows for the herring."

The Ode Camp on Newcastle Island was no longer in use by the time Ken Hamagami worked the 1941 herring season at his father's saltery on Galiano Island. Ken explains that he didn't know all the details about Matsuyama's involvement with the camp. He says his father's birth name was Ode, but he had agreed to take the name of a childless couple in Japan so that their name, Hamagami, would live on. In Canada, he continued with the name but gave his herring saltery business his original name, Ode. Probably Matsuyama had purchased the first Ode Camp by the time Randy Thompson was fishing.

Ken Hamagami's wife, Jean (Hideko), is the daughter of his father's friend and fellow saltery owner, Rikimatsu Tabata. Sometime in the late 1920s, the Tabata saltery moved down from Jessie Island in Departure Bay to occupy a site on the southwest side of Galiano Island, next to the Ode camp. There were six salteries on Galiano in 1941: Sugimoto, Tanaka, Shiraishi, Yip, Tabata and Ode.

Ted Nakatsu worked at the Yip camp in Otter Bay on Mayne Island in the late 1930s. He explains that at that time Matsuyama was also involved with the Yip Sang plant. Ted had a job salting chum salmon for Matsuyama, at Jedway in the Queen Charlottes. At the end of the season, he was offered a job at the operation Matsuyama ran with the Yip family. He came down and fished there on the *Merle C* and one of the fleet of *Yip*

boats. He recalls the making of a set. "I think there was fourteen to sixteen on each boat. They had to be half white and half Japanese. The net was laced together between the two boats. I don't know how they tied the cork line. It must have been simple, because when it blows, all of sudden we have to split it and run for the harbour.

"We fished at night. There was a scout out ahead in a skiff with a flashlight. He just looked for bubbles, or sometimes there was a jumper. He didn't have a feeler wire. Sometimes we looked for hours and can't find it, so we go in and sleep, then go out again just before daybreak. Most of our fishing was done in Swanson Channel. Ode and Tabata fished around Porlier Pass. The skippers stayed in the wheelhouse on the seine boats. The scout was the boss. With a flashlight, he signals the skipper to make a set, and a lot of yelling I remember between the scout and the skippers. The scout makes all the decisions.

"We set full bore. There was a line tying the two boats' bowposts together. There was a big post with a metal piece passed through. It was my job to let that rope go when we start a set. I remember one time it started slipping and scared the shit out of me. The boats went out from

One of several boats with the *Jessie Island* name. At least five of the fleet were still registered in 1992. Some were built in Steveston while others, including one that was later renamed the *San Jose*, were built at Ritherton Bay in Barkley Sound. (Hamagami family)

each other in a big circle. They towed a little while, and then the bows came together and the rope was thrown over. It was my job to tie them together before the crews started pursing one line on each boat. I start to take up the slack as the boats come together. After we get the rings up we start to pull the net with a song. It's like, when there are more than one pulling, you've got to have something to co-ordinate the strength. I bet even now there are white people who used to work on that who still remember that singing.

"The Fisheries used to come and measure our net's length, and every time they go, the crew used to extend the net. When the herring are dried up in the net, three or four tender boats put scows onto the cork line. We had lots of tenders because the season starts after salmon season in October. We also had small, packer-sized towboats to help. We brailed pretty fast. At first, the scow is so high it is hard to brail. But after ten or fifteen minutes, we get enough into the scow that it tips. Then it is fast. The brailers are an oblong frame with two handles and two guys. I bet those things hold two or three hundred pounds. If there is one hundred tons we would have about two scows with six or seven brailers on each. We never put fish on the boat. Inside the hatch there is ballast on one side so the boat won't tip with the weight of the fish in the net. When you are empty the boats lean away from each other."

Randy Thompson described the double-pursing of the net. "When we come together they had a thick rope with two eyes on it that went from the tow bit on one boat across to the other one. Then we purse one end of the purse line on each boat. The last piece of lead line on each boat was made with enough space between the ring straps or stoppers to reach the length of the two boats. Then we would pick up that piece of lead line and hang it on the two boats so that it wasn't hanging down there for the herring to go through. Each boat brought a part of the lead line up. Then you would undo the purseline in the middle and take it out of the rings and pull the net by hand."

"The Japanese and whites slept separate," Ted Nakatsu explained. "I think half slept on each boat. Once in a while we worked in the saltery. There were big canvas bags the size of a room. The herring are put in there with salt and left for a few days. They put them on the floor and put it into boxes. I think they were 500-pound boxes."

Randy Thompson also recalled that on one of Kasho's boats the whites slept down in the fish hold. They had bunks around the sides and a cookstove, for which they installed the pipe before lighting and ran it up through the open hatch.

The late Louie Percich hadn't been long in Canada when he went fishing for Kasho on Galiano Island. As Louie recalled it, the Europeans and Japanese may have slept separately but they shared the work with a song and made fun of the Fisheries officers charged with enforcing the racist regulations. "The crew was supposed to be only half Japanese. One day the Fisheries inspector came when we were tied up. He wanted all the white fellows on one side and the other fellows on the other side. There were two natives from Nanaimo. They lined up with the Japanese and the inspector said, 'What the Hell are you doing there!'

"The natives didn't know if they were white men or what. I still laugh when I think about it. They were good to work with. We sang in Japanese when we worked. We were paid twenty-five dollars a month. In the day we worked salting in the plant and at night we fished.

"When we were ready to set we laced the net together. There was a big pile on each boat. The boats go out in a circle then meet and tie the bow together and purse from both boats. When you get the rings up you pull net from both boats. Then you come to the bunt in the middle. When we get to the fish in the bunt we tie a scow along the cork line. You sit on the side of the scow with one leg in the water and the other one in the scow. You have these brailers that you work by hand with two men on each brailer. It's hard work. Jesus, it's hard work. Pulling and singing all the time. There was no echo sounder in those days so we snagged a lot and I did a lot of net work."

Jean (Tabata) Hamagami grew up in Vancouver but she recalls visiting the Jessie Island saltery of her father, Rikimatsu Tabata, when she was a small child. The Tabatas had a house higher up on the island, but

Many herring salteries also salted dog salmon and their eggs for export. This may explain the barrels at this Gulf Island saltery where a freighter is loading cargo. (Hamagami family)

she remembers seeing her father going out in the scouting skiff to guide the boats to the fish. Ken Hamagami says his dad, who co-owned the saltery, had a big blue speedboat called the *Blue Jay* which he used for a variety of purposes. This may have been his scouting boat.

It was a good life for a seventeen-year-old boy, taking a year off school with plans to return to business school and eventually to help manage his father's extensive holdings. It was 1941, and the herring season, which started in mid-October, ended December 6 that year. The Ode camp held a party for all the fishermen, saltery workers and the crew of a firm that had leased space to can herring. There had been a dance and a little drinking, all made lively by the presence of a number of young women who worked in the saltery. Afterwards it was speculated that a young couple may have wandered out to the end of the wharf and left a smouldering cigarette butt on one of the tarred cotton seines piled there. Whatever the cause, a fire started on the outside of the dock and, fanned by strong winds, swept shoreward through the saltery, up to the fishermen's bunkhouses, into the Chinese bunkhouse, the women's quarters, the Indian rooms, the store and office, then the watchman's house and finally to owner Yasaburo Hamagami's house. The tide was low, and the location had no water. There was nothing the people could do but stand back and watch. It was said later that the glow was seen in Steveston. Finally the Ode Camp people went next door to the Tabata Camp to get some sleep.

The Galiano Island herring salteries, the site of the December 6, 1941 fire that marked the end of an era.

December 7, 1941 dawned on a depressing scene for young Ken Hamagami. Where a fine, big saltery had been, only ashes remained. Everything had burned. But if this seemed a catastrophe, it was dwarfed by events later in the day, when the radio carried word of the Japanese attack on Pearl Harbor and spelled the end of all the work and dreams of immigrant and Canadian-born people of Japanese descent.

SEINING FOR PILCHARD

"**I** was thinking to build a special boat for pilchards. I figured this out. When you're outside with a fairly flat-bottomed boat she goes up on a wave and pounds down and makes a noise that scares fish. So you build a boat like a yacht so that she cuts through the water. So that's the way I built the *Western Challenger*."

Charlie Clarke had several boats built over the years. He viewed their design rather as a fly-fishing sports angler sees the design of a new fly. The challenge is to understand not just the fish, but its environment: where it will feed, the patterns of its travel, how it will behave in varying tides and weather conditions, how it schools in different light or at different times of day and on through a myriad of variables. Some theories are proven in practice and become the norm for the fleet, while others are implemented with much anticipation and quietly let go as they fail to result in more fish.

When Charlie had the *Western Challenger* built in 1941 the pilchard stocks off the west coast of Vancouver Island already were showing signs of decline. The pilchard is the well-known sardine of California. It comes to British Columbia only at the limit of its northern migration. The first recorded appearance of pilchard in BC waters was in 1900, but by 1929, 81,250 tons were reduced to oil and meal, 4,900 tons were canned and 150 tons were used as bait. These products had a combined value of $2.2 million, and over $3 million had been invested in boats and plant equipment for the fishery. Pilchard oil, by far the most valuable product, was used in soap, paint, margarine and shortening, while the meal went into animal feeds.

The runs varied in the extent of their northern migration, occasionally ranging as far north as the Queen Charlotte Islands and into Georgia and Queen Charlotte straits. In the early years, pilchard was taken in the sheltered waters of the inlets on the west coast of Vancouver Island, but by the 1940s the seine fleets were going farther offshore and south to international waters three miles off the US coast. The first recorded commercial catch was for 80 tons in 1917. The catch grew quickly to 4400 tons in 1920. It then hovered around 1000 tons before skyrocketing to nearly 16,000 tons in 1925, 48,000 in 1926 and 68,000 in 1927, and peaking at 86,000 in 1929. In that same year the catch in California, where the pilchard spawned, was 325,000 tons.

The result of such fishing pressure was inevitable. By 1946 the catch fell below 4000 tons, and in 1947 it was a mere 500 tons. Many

fishermen maintained optimistically that it was just a cyclical phenomenon and the pilchard would return. But the fishery was over. By the 1950s newly built pilchard boats from Washington State were being offered for sale to Canadian salmon fishermen at bargain prices.

This dismal management record was not in Peter Wallace's mind in 1926 when he sold his interest in the Kildonan cannery in Barkley Sound to BC Packers. He went north to Kyuquot and purchased the whaling station there, complete with two of the big iron whalers. He built a reduction plant with the expectation that the summer pilchard fishery, combined with a winter herring fishery, would provide year-round supplies of fish.

Charlie Clarke recalls: "When Peter Wallace sold out and quit at Kildonan he asked if I'd go with him to Kyuquot. I thought it over and said, 'OK, I'll go with you.' He wanted another good skipper so I got Johnny Watson's dad. They had two old whalers and called me down to Victoria to have a look at them. He said, 'What do you think? Will they make seine boats?'

"I said, 'No, they'll never make seine boats. They're too narrow in the stern. There's no place to put a seine table for the net.'

"He got somebody else to look at them and they decided to put the table in the centre of the boat and run the net out from there. In them days we didn't have very long nets, only 200 fathoms. If you turned your wheel hard over you could get back to the end of your net. But if you straightened her at all you were lapped back and couldn't get to your skiff.

"The first time out Peter Wallace went along with me. We made a

Pilchard came in very large schools and were usually loaded onto scows as a small seine boat could only hold twenty or so tons. In this early photo, the *W. No. 1* seems to have more fish than wanted as the crew are hauling the web up by hand to spill some fish over the cork line. It may be that the net was wrapped around the boat and this was the only way to get it straightened out. Note the canvas skirt placed over the stern, probably to act as a net guard around the rudder and propeller. (PABC-C6079)

set. When we started to purse, the net was sunk at the centre. Somebody hollered, 'You snagged! You snagged bottom!'

"I said, 'You can't snag in eighty fathoms of water. I'm darn sure it doesn't take bottom here.' Now the net went out over a roller on the side of the boat just like a big rope going through a block. What happened was that the lead line had gone down one side. When half of the net was out the lead line went over the other side of the cork line. So we had to go out in the skiff and pull the damn thing all up and put it back on one side.

"When we finished, I said, 'Mr. Wallace, that won't work.'

"He said, 'You've got to make it work.'

"So a few of us got together and thought of keeping the seine table in the middle of the boat where they had it and putting a pin in the corner of the table so you could swing the whole thing out over the side of the boat. It would be not exactly square but kitty-corner. Then we put railroad iron underneath with hooks so that it couldn't upset and we had a stopper on it so that it could only go so far. I think I caught about 800 tons with it, but still there's no way to brail into the boat. There was so much deckhouse in the centre of the boat. Where the captain and officers had a wardroom aft, that was all taken and gutted out and made into a fish hold. Well how the hell are you going to get fish into it?

"They built a big hopper up in the sky to dump the fish back in there. Well how are you going to get on the outside with the bloody brailer going back? You couldn't hit the bloody thing and if you did you'd knock the wheelhouse off. So that didn't work.

"One time we got out and the boat got turned around. We were all by ourselves and the net was in the water wrapped around the boat. We had set at about five o'clock in the evening. About three in the morning I saw a troller coming. I flashed a light and blew a whistle at him. He came over and I said, 'Would you turn us around so I can get the net pursed?' So he did that and I said, 'I'll give you ten dollars.'

"He said, 'No, I don't want that, Charlie.'

"But I gave him the ten dollars anyway and we got the net back. In the meantime we had caught pilchards and the dogfish had caught us and just riddled that net from stem to stern. We worked all night mending the damn thing. On the way in that morning my crew said, 'No more of this. You go up and tell Mr. Wallace we want a guarantee of $2000 for the season or get us a real seine boat.'

"He had a seine boat there, the *Maid of Eacool*, but he wouldn't let us use it. So Mr. Wallace said, 'Dash it! Dash it!' That's the way he talked. 'If we give you this seine boat none of that dash fishing during the night!'

"Then I got mad. I told him what had happened. So we got the seine boat. There were a lot of pilchards inside that year. Kyuquot Sound was full of them. Some of the inlets, like Deep Inlet, we couldn't fish in there for dogfish, they would eat your net up. But we put in about 3000 tons of pilchard and we used those whalers as packers towing scows and that's how we finished the season."

Innovation has been a constant challenge and delight to seine fishermen as technology is mated with experience to provide greater productivity. But plans imposed from shore-based management have generally been met with scant enthusiasm. Charlie's story illustrates an important lesson that all successful processors learned, or suffered the consequences: Don't tell a good fisher how to catch fish. The boats and gear that evolved in British Columbia grew almost entirely from ideas borrowed from other locales, or from the minds and experience of the fishers themselves. These people were not factory workers prepared to run a machine for a few hours in return for a wage. Charlie's crew demanded $2000 to stay on the whaler not because they wanted a wage, but because they wanted a decent seine boat.

By the end of the 1928 season, Charlie had had enough of other people's boats. Licences were more readily available to individual white fishermen and Charlie went off to Vancouver to shop for a boat. It was a good time for an ambitious young man to be looking for a seine boat. They were well known as money makers, and there was a lot of loose money around the Vancouver waterfront looking for a good investment. This was the height of the rumrunning era and Charlie borrowed $10,000 from liquor magnate Henry Reifel. He found the *Port Essington*, a seine boat with a registered length of 57 feet with a 16-foot beam. It had been built in 1927 by S. Yamanaka of Steveston for Hichisaburo Kameda. Since seine licences were so restricted for Japanese Canadians, it is likely that Kameda was not able to fish his new boat and sold it as a result.

That first year Charlie nearly lost the boat himself. "My charter to the company that year [the 1928 fishing season] would be $10,000, so I told them to pay it to the bank every month so I wouldn't have to pay too much interest. But when I came back down to Vancouver at the end of the season they hadn't paid. They paid two months, that was all. So I lost about $6,000. I went to the company and said, 'Give me one of the seine nets for my part. That will clear me up.'

The *Port Essington* was built for Hichisaburo Kameda in 1927 by Yamanaka of Steveston. Charlie Clarke bought her with money borrowed from liquor magnate Henry Reifel. Her name has since been changed to the *Pacific Marl* and she was still working in 1993.

Ritchie Nelson began packing troll-caught salmon from the west coast of Vancouver Island to Seattle with the *Gradac,* which was built in the US in 1915. He went on to build a large fleet of green-hulled seiners. The fleet and canneries were sold to BC Packers in the 1960s. The *Gradac* is still working.
(VPL 25074)

"But oh no! They wouldn't do that. So I went to the banker and he said, 'Oh, don't worry about it, Charlie. You're young, you'll fish again next year.'"

That next year Charlie found his niche with a young entrepreneur named Ritchie Nelson. Ritchie, the son of a Norwegian immigrant who died when Ritchie was an infant, grew up gillnetting on the Fraser. While still in school he bought his first boat, a 5-hp gillnetter. In 1917 he went halibut fishing on his brother Norman's newly acquired 40-footer, the *Bayview.* In 1919 they took the boat down to the west coast of Vancouver Island and began cash-buying troll-caught salmon for resale in the US. By 1920 the Nelson Brothers had bought an American seine boat, the *Gradac.* Then, in 1924, they had the first diesel-powered Canadian seine boat, the *Sundown,* built at Menchions Shipyards. The Nelsons may have fished with the seiners *Gradac* and *Sundown,* but they were primarily used to pack troll-caught salmon until the trollers formed a co-operative to provide their own packing.

In 1930 Ritchie was asked by the manager of the Canadian Packing Corporation, a subsidiary of the giant California Packing (Delmonte) Corporation, to manage the fishing operations for the pilchard reduction plant at Ceepeecee, near Tahsis. At that time pilchard oil was selling for 25 cents per gallon, but the CPC head office decided to hold its stock for a higher price. When the price dropped they lost heart in their Canadian subsidiary and offered the million-dollar operation for $50,000 to the Nelson Brothers, who negotiated an even lower price. Nelson Brothers now had a plant and four seine boats, including the *Luac, Ribac* and *Amalac.*

From that date to his retirement in 1966, six years after he sold out to BC Packers, Ritchie Nelson proved to be a master at the art of matching fishers with boats. Over the years his company owned many boats but his preference was always for the owner-skipper who had a vested interest in caring for the vessel. Charlie Clarke quickly paid for the *Port Essington* while fishing for Ceepeecee, and when Ritchie bought the plant he put Charlie in charge of the fleet. Ritchie soon displayed another technique for keeping good skippers—encourage them to buy new boats. If skippers owe the company money, they won't be tempted away by offers from competing companies.

Charlie recalls Ritchie coming to him one day in 1936 and proposing that he build another boat, a bigger one. Charlie shopped around and found Joe Tara out in Ladner. "He was the cheapest. He'd build me one for $21,000. A 63-footer, 20 feet wide, the *Western Monarch*. We put a Washington Iron Works engine in her. I was going to get an Atlas but they wouldn't promise me the engine in time to have the boat ready for the season. It was a clutch job."

Although it did have a top wheelhouse added years later, Charlie had the new boat built without one. "I didn't put a top wheelhouse on," Charlie explained, "because when you're looking for pilchards you can go over a spot of fish and not see them. They just flip on top of the water now and again. When you have a top wheelhouse you can't see behind you. We just had a bridge up top and we sat up there and could look all around the horizon. I always had a man on each side of me and one up the mast in a barrel."

Charlie fished the *Western Monarch* for several years. Then, shortly after the beginning of World War Two, Ritchie said, "Why don't you build another boat?"

Experience with the *Western Monarch* with her 20-foot beam, and increasing offshore searches for pilchard, led Charlie to new ideas. "That's

The *Western Monarch* was built for pilchard fishing in 1937 with an open top so that Charlie could have unlimited visibility while searching for fish.

the 73-footer that I built like a yacht, with the narrow bow coming out to an 18-foot beam. When I was building the *Western Challenger* the Fairbanks engine man came to me and said, 'I tell you what, you're always kicking about the noisy clutches in the engine. We've got a new 6-cylinder engine with a real quiet clutch. The engine is a 2-cyle type rated 180 hp at 450 rpm. We'd like you to try it, so we'll give it to you at cost price.'

"It seemed like a good idea. You've got to have a quiet clutch with no noise to scare the fish. I knew from the whalers that even the propellers make a noise. They bored them and leaded them so there is no ring to the propellers. I did that to all my boats so that I could crawl up on fish with no trouble. This is true for pilchards, herring and salmon. On salmon you use a plunger on the end of a pole to make a noise on the water to scare fish away from the end of the net.

"So I built the boat with the quiet clutch and the new Fairbanks engine in 1941. We got out fishing and the clutch was fine. The engine was a 2-cycle diesel and we were going to Prince Rupert with the seine on.

Pretty soon fire started blowing out of the smokestack. We had to get the hose and put water on the seine all the way to Rupert. We burned ninety gallons of lube oil on that trip. So when we came back I told them about this. They sent a crew from down the States and tore the engine all down and put something in so as to fix it up. But still a 2-cycle won't run slow without building up carbon. On a seine boat you have to idle most of the day while you wait for fish or work the net. I couldn't fish the boat. As soon as we would idle she would build up and then out she'd go and burn up the stack again.

"The co-op was looking for a fish packer and they came up to the company and asked if they had any boats for sale so I offered them mine. I told them all about it and that it ran good full speed but it won't idle so it's no good for seining. So I sold that boat for exactly the same price I paid for it, $32,000. They renamed her the *Co-operator IV*. Years later they sold her and she became the *White Swan*. She has a new engine now and is a good seine boat."

Charlie's progression from the *Port Essington* to the *Western Monarch* and then the *Western Challenger* was mirrored in the experience of a number of other fishers and builders throughout the late 1920s and the 1930s. The vessels became bigger, with 72 feet becoming a norm for large boats, mainly due to a boatbuilding subsidy program. By World War Two virtually all new seine boats were diesel-powered and a good many older

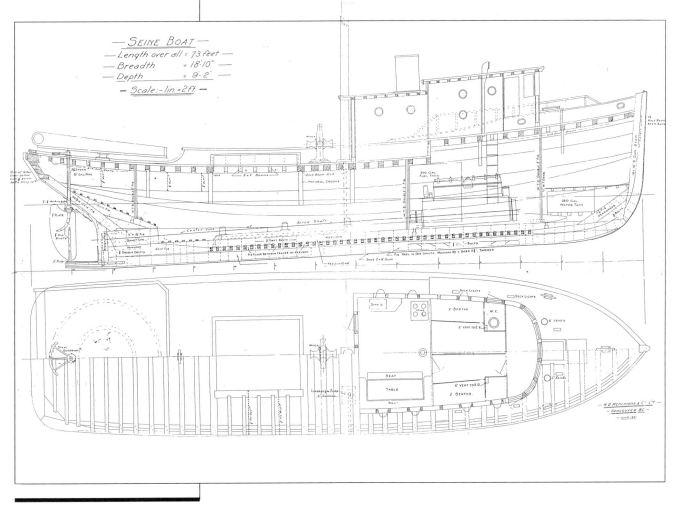

The *Western Challenger* became the packer *Co-Operator IV* and then the dragger *White Swan*. She is back to seining now, and another boat has the name *Western Challenger*. In Canada only one boat can be registered with a given name.

boats had been repowered with low rpm engines. When the pilchard ceased their annual appearance on the BC coast in 1947, the boats of the pilchard fleet became the vanguard of the San Juan outside salmon seine fishery, and a fleet of Canadian boats that developed the Bering Sea halibut grounds in a longline fishery.

MENCHIONS: LAST OF THE COAL HARBOUR YARDS

During the 1920s and 1930s Vancouver's Coal Harbour was the heart of the boatbuilding industry on the West Coast. Harbour, Stanley Park, Turner, Fenner & Hood, Union, Hoffar–Beeching, Boeing and Menchions—they all had yards in Coal Harbour where they built sturdy wooden boats for fishers, towboaters and anyone else who had business up along the BC coast. The yards are gone now, victims of escalating land values, but many of their boats are still in service. The seine boat *Zeballos* is a fine example. Built in 1918 at W.R. Menchions Shipyards for Pete Anderson, she was still fishing in 1992, two years after the Menchions yard itself had closed. At seventy-five she is probably the oldest surviving Menchions boat, but there are many others still in service that have passed the half-century mark. Obviously, Menchions made boats to last.

W.R. "Bill" Menchions, founder of the yard, was born in 1871 at Bay Roberts, Newfoundland, where it was said men would go into the woods in the fall and come back out in the spring with a newly built boat. His family came from seafaring stock, emigrants to Newfoundland from the Isle of Guernsey in the 1700s.

The *Zeballos* has had many owners since her launch in 1918. Tom Boroevich (above; AHB) owned her from about 1950 until 1971. Now she is retired so that a new aluminum seiner can be built for her licence.

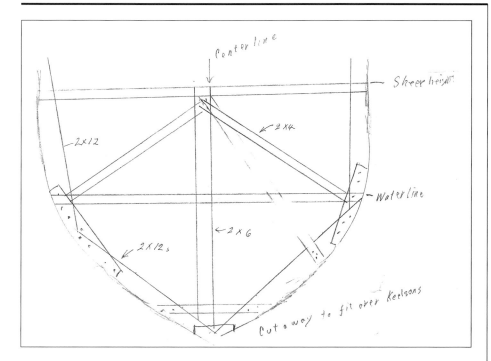

Centerline

Sheer height

2×12

2×4

Waterline

2×6

2×12s

Cut away to fit over Keelsons

A wooden boat is a sum of many unique individual parts. Bert Menchions sketched the making of a station mould for a bent frame boat to show how precise the measure must be even on parts that will not be in the final boat. Repairs require the precise duplication of rotted or broken parts that often no longer have their original shape. For the last twenty years of its life the Menchions Yard did only repairs to wooden boats. Below, Chi (Richard) Sheng Chen fabricates the lower knee that will hold a new horn-timber on the forty-year-old seiner *Nesto* in 1988. The job cost about $25,000. Even allowing for inflation, this represented a good part of the original cost of the boat. When Menchions closed, Chi took his skills to the Musqueam Indian Band's Celtic Shipyard where he was still nurturing a few more years out of the old boats in 1993. (AHB)

As a young man Bill Menchions fished and sealed off the coast of Labrador. In the late 1890s he spent two years in the Yukon building paddle wheelers before settling in Vancouver, where he worked for Andy Wallace, founder of Burrard Shipyards, then started on his own in 1909. The Easthope Brothers had built a little yard behind their engine shop on Coal Harbour. It was this yard that Bill took over and began installing Easthope marine engines in 24-foot Columbia River-type gillnet boats.

By 1925 Bill's reputation was well established and earned him a contract to build five pilchard seiners for Wallace Fisheries' Kildonan Plant. This was the plant and the firm for which Charlie Clarke had started working, the same people who had brought in that first fleet of seiners from the US. The new boats represented the next generation of BC seine boats. The contract specified that each of them would have an overall length of 62 feet, a beam of 15.5 feet and a draft of 5.5 feet. The specified materials reflected the availability of fine old-growth lumber. The 10- by 11-inch keel and the 6- by 12-inch keelson were select fir in one length. Spotted gum wood was used for the 10-inch stem, the 10- by 12-inch stern post and the 2- by 10-inch shoe. Frames and floors were 2- by 3-inch oak. The frames were spaced on 10-inch centres except under the engine bed where they were on 6-inch centres. Planking was 1 3/4-inch fir in long lengths, clear and free from sap.

Each of the five vessels had accommodation for six men in the fo'c'sle and one in the pilothouse, where doors and windows were made of teak with copper drain trays under the windows. The engine, dynamo (generator), batteries and seine winch were supplied by the owner.

Nearly all of the wooden seine boats built in British Columbia have had planks nailed to bent oak frames. The Wallace boats were built to this method where a number of cross-sections of the hull's dimensions were drawn full size on the loft floor in a process called, appropriately, lofting. Once the sections were drawn on the floor, molds were made with rough lumber, but to very precise dimensions on the outside to define the

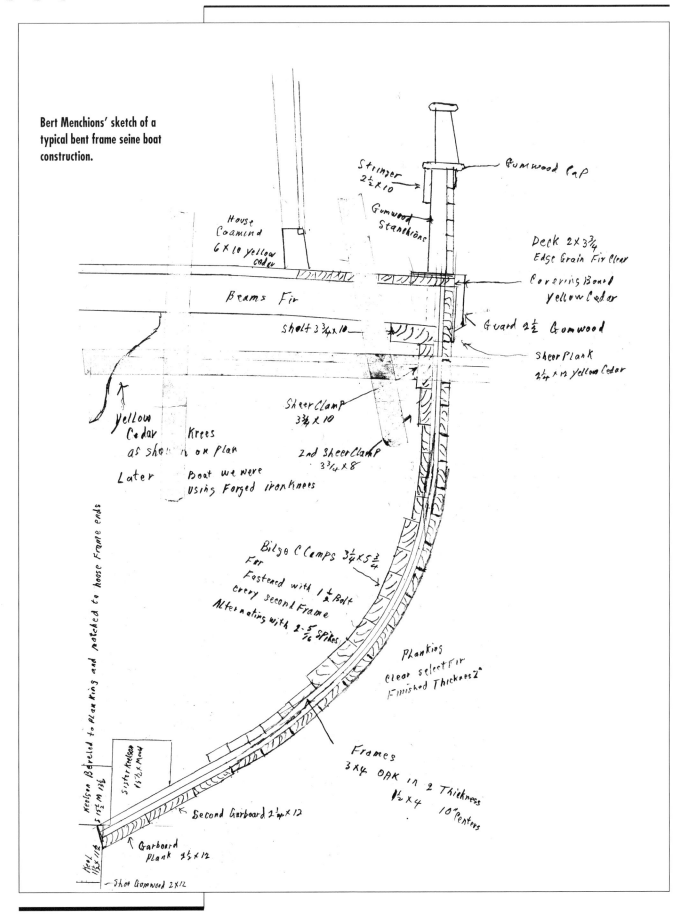

Bert Menchions' sketch of a typical bent frame seine boat construction.

Stringer 2½ x 10

Gumwood Cap

Gumwood Stanchions

House Coaming 6 x 10 yellow Cedar

Deck 2 x 3¾ Edge Grain Fir Clear

Covering Board Yellow Cedar

Beams Fir

Sheet 3¾ x 10

Guard 2½ Gumwood

Sheer Plank 2¼ x 12 yellow Cedar

Sheer Clamp 3¾ x 10

Yellow Cedar Knees as shown on plan

Later Boat we were using Forged iron knees

2nd Sheer Clamp 3¾ x 8

Bilge Clamps 3¼ x 5¾ Fir Fastened with 1½ Bolt every second Frame Alternating with 2-⁵⁄₁₆ Spikes

Planking clear select Fir Finished Thickness 2"

Frames 3 x 4 OAK in 2 Thickness 1½ x 4 10" centers

Keelson Beveled to planking and notched to hoose Frame ends

Keelson 6½ x M 13½

Sister Keelson 1½ x Mood

Second Garboard 2¼ x 12

Garboard Plank 2½ x 12

Keel 1½ x 14

Shoe Gumwood 2 x 12

hull's shape. These molds were then set up vertically on the keel and keelsons at each station, plumbed for vertical, braced in place and joined with longitudinal ribbands of 2- by 3-inch lumber. A seine boat typically had around ten of these molds about 6 feet apart. With the ribbands in place, the steamed oak frames were bent into place with the bottom end fitted into sockets that were cut into the keelson and the upper parts clamped to the ribbands. The frame was fastened to the ribband with common nails and the clamps removed.

With the molds and the frames in place, work started on the sheare clamp and deck shelf. These timbers, together with some deck timbers, helped hold the hull's shape. Parts of the molds were cut away to allow the bilge clamps—large dimension planks running the full length of the vessel—to be bolted to the frames at the curve of the bilge. Planking started with the sheare plank at the top of the hull and the garboard next to the keel. The ribbands were removed as the planks went on. In their turn the molds were removed from the planked hull.

The stern was defined by a series of heavy timbers known collectively as the deadwoods. The deadwoods were mounted on the after part of the keel and included the horizontal shaft log through which the propeller shaft passed before reaching the vertical stern post. Mounted on the shaft log was the horn-timber which extended upward, just as its name suggests, to form the deadrise, the sweep that carried a boat's hull up over the propeller to a point where, on most seine boats, it supported a nearly square, heavy-timbered stern.

The Wallace vessels cost $6400 each; or, allowing a twenty-fold increase for inflation, $128,000 in 1993 dollars. Even if the materials were available, a price of three to four times this amount would be conservative today.

Menchions launched all five boats in 1926, and Peter Wallace named them, consecutively, the *W No. 7* through to the *W No. 11*. Big boats for their day, they were built to take the weather of the West Coast and, if necessary, pack a real payload of fish, though scows were still used for much of the pilchard and herring packing. The boats fished only two years for Wallace Fisheries before the firm was bought out by BC Packers. Over the years they were eclipsed in size many times, but the fact that all five of these boats were still working in 1992 is testimony to the quality of the craftsmanship at Menchions Shipyards.

The 1920s were great years for all of the yards along the Coal

The loss of the old growth forest was felt by boatbuilders as early as the 1960s when they could no longer get the select lumber described on this 1951 invoice for $545.71. By the 1980s such lumber couldn't be had in any quantity at any price. Shipwrights today routinely find that planks replaced five years ago will be rotting out while original wood from decades ago with its straight, tight grain will still be sound.

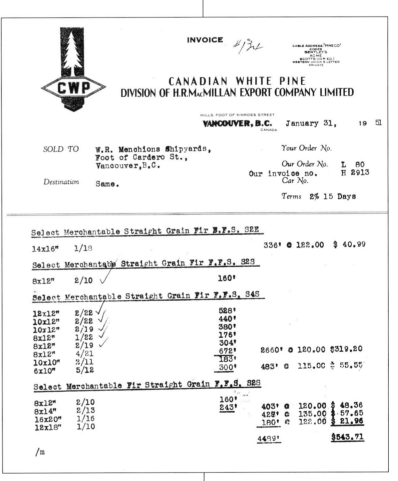

The original Celtic Shipyard was operated by BC Packers at the foot of Blenheim on the Vancouver side of the Fraser North Arm. In this 1920s photo, several of the Wallace Fisheries boats are tied at the yard, having recently been acquired when BC Packers bought out Wallace. Note that the early dodgers were often canvas-covered pipe. The *W. No. 10* has her canvas up while the *W. No. 11* has removed hers. (VPL 10277)

Bert Menchions recalls taking a pay cut to get the contract to build the *Meander* for George Kidd. She is sometimes used as a charter boat in Vancouver today. (VPL 15923)

George Menchions, Bert's son, operated the yard until the foreshore lease ran out and the waterfront was slated for "yuppification." The yard equipment was auctioned in September 1990, marking the end of Coal Harbour as a working waterfront. (AHB)

Harbour waterfront, and not only for seine boats. Prohibition in the US created a demand for specially built rumrunners, and the logging industry needed towboats to move the booms from upcoast to the Vancouver sawmills. Lumber barons and other capitalists were reaping huge profits from the resources of the province, and they wanted private yachts. Menchions, for example, launched the 132-foot *Norsal* in 1922 for N.R. Lang of the Powell River Company. As well, government departments and churches needed boats to take their representatives up the coast on business. In the 1930s things slowed a little, but there was still a demand for yachts from people like George Kidd, president of BC Electric, for whom Menchions built the *Meander* in 1934. The yard crew agreed to a pay cut to 5 dollars per day to help Bill Menchions get that job.

The war years created a demand for seine boats that continued after the war. In 1946, when Bill Menchions died, his nephew Bert took over full responsibility for the yard. About this time he built three boats with sawn yellow cedar frames. Yellow cedar, or cypress, has a very special place in BC boatbuilding. Soft and easily worked, it is lightweight and, most important, contains a resin that makes it remarkably rot-resistant. Quite rare in the early days of the century, it was in common use by the 1940s for stern timbers, deck beams, and bulwarks. Now Bert took it a step farther and used it for the frames, or ribs, of the three new boats.

Yellow cedar cannot be steamed and bent like oak, but oak can only be steamed and bent up to a certain dimension. Shortly before his death in 1985, Bert explained in a letter the importance of sawn yellow cedar frames in two of Menchions' boats. "Sawn frames were used when they built the big tugboats that used to be famous on the coast. Like the *Sea Lion*, the *Ivanhoe*, the *Master* and the *Prosperative*. Those boats used fir frames. I think for one thing it would be almost impossible to get good [bending] oak in the length that would be needed for those big deep tugboats. The longest oak we used for frames was about sixteen feet, with maybe a few of 18 feet, and that is quite a sized chunk of clear oak."

Sawn frames could also be made heavier than bending oak, which is limited by the builder's ability to bend it into place. Bert explained that when an architect prepares a set of lines for a boat, the dimensions are to the outside of the planked hull. When the shipwrights are drawing a full-size cross-section on the loft floor to be used in making the ten or twelve station molds on a bent frame vessel, or the seventy or more patterns for a sawn frame boat, it is necessary to allow for the plank thickness before drawing the shape on the floor. This is just one of many calculations that have to be done more frequently with sawn than with bent frame construction. The frames are not sawn square; they must be shaped or bevelled to the round of the hull.

"Not only that," wrote Bert, "the bevel for each frame changes a lot for the whole length of the frame . . . from top to bottom. Also the joints have to be marked on the pattern to suit the width of the material that can be used. So you see it is quite a lot of work. I, being the loftsman, had really sore knees by the time I finished laying out a sawn frame type of boat."

Two of the new boats, the *Kitimat*, a Fisheries patrol boat, and the *Cancolim*, a launch for the American Can Company, were designed by Halliday, the Seattle naval architect. The third was a 73-foot double-decker seiner, the *Western Warrior*, designed by Robert F. Allan. Other boats were built at Menchions after the *Western Warrior* was launched in 1944, but judging by Bert Menchions' enthusiasm for the sawn frame boats, it may be that the pinnacle of seine boat construction on Coal Harbour is carried in that hull with its very West Coast heart of yellow cedar.

Menchions built boats for most of the major fishing companies, including the *C.F. Todd* for J.H. Todd & Sons in 1940. She is a good big boat with a captain's stateroom between the wheelhouse and galley and a full bulwark forward to give her a nice shear line.

The *Western Warrior* fished the 1993 season and shows every sign of continuing to work for years to come. As such, she could also stand to represent the many yards that produced fine wooden boats along Vancouver's Coal Harbour waterfront. Some of the yards were short-lived with small production, while others, like Benson's, have many vessels still working the coast. Land values combined with residential and recreational pressures to drive industry from Coal Harbour, but it was labour costs and lack of good wood that ended the wooden boat era. When George Menchions closed the yard which he was the third generation to manage, it hadn't built a boat in nearly twenty years. Other yards switched to steel before closing or moving to other locations. In 1993 a group of individuals who had operated businesses or worked there arranged to have their names engraved on a ship's bell to commemorate this dynamic and productive few blocks of waterfront. But the greatest testament to the skills of these workers will always be the boats they created.

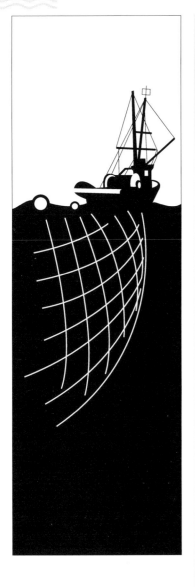

THE SMALL ILLUSTRIOUS LEGACY OF K.M. BOAT WORKS

Not all Coal Harbour shipyards were long-lived or prolific. Some left a legacy based on the quality of the few boats that they did produce. One such yard was the K.M. Boat Works. Named for the owners, Sajiemon Kuramoto and Jitsuji Madokoro, the yard operated from about 1937 until the internment of Japanese Canadians at the beginning of 1942. The yard was the culmination of many years of work by the two owners. Madokoro's father had come to Canada before the turn of the century, but returned to Japan where Jitsuji was born in 1893. Jitsuji came to Canada himself in 1910 and met up with Sajiemon Kuramoto, who was from the same Japanese village. Kuramoto had been in Canada since 1900. Both men worked in a variety of places and at various jobs. By the early 1930s Madokoro was working for Mr. Hisaoka at Stanley Park Boat Yard, and Kuramoto was at another Coal Harbour shipyard, the Bidwell Boat Works, later known as Union Boat Works under the ownership of Tom Nakamoto. In 1937 the two men opened K.M. Boat Works on leased land between Bidwell Street and Stanley Park. When the lease ran out, they moved briefly to the foot of Victoria Drive, but their tools were constantly stolen so they returned to Coal Harbour. This time they leased land near the Millerd cannery just behind the old ice rink at Denman and Georgia.

The Kuramoto family lived in a house belonging to the Millerd cannery, while the Madokoros had a house at 1170 Powell Street. The Vancouver Ship's Registry showed several fish boats built by K.M., including: in 1937 the 31-foot *Flora L* for David Lornie and the 27-foot *Aonoa* for Arthur Harrington; in 1940 the *Rosalie I*, a 31-foot troller, for the Ryalls; and in 1941 the 36-foot *Muriel D II* for John Daly, who later owned the *MoreKelp*, the 35-foot *Crown* for Yoshio Nadokoro of Tofino, and the 36-foot *Lake Biwa No.2* for Yoshimi Nobuoka. Only one seine boat was listed: the *Great Northern 5* was built at K.M. Boat Works and documented on September 5, 1940, with a registered length of 61.2 feet. It was built for the owner of Stone Bros. Towing in Port Alberni.

George Kuramoto of Prince Rupert, a retired fisherman and son of Sajiemon, remembers his dad buying most of their

Jitsuji and Hiroye Madokoro in the 1940s. (Madokoro family)

K.M. BOAT WORKS

hardwoods from Fyfe-Smith, the hardwood specialist, and that they could also get really good edge-grain fir in those days. An indication of the care and time that the builders put into their boats is given by Jitsuji's son, Mamoru, who recalls: "When I was a kid we used to go to Surrey, to Strawberry Hill to get natural crooks. We knew a fellow named Hashimoto who owned land there."

In 1942 the Madokoros and Kuramotos were sent to the Greenwood Camp near Grand Forks and were not allowed back on the coast until 1949. "After the war Dad built with Nakade for a while. Then he went to Sunnyside [on the Nass River]," Mamoru remembers.

Sajiemon and Yoshie Kuramoto at Greenwood Camp. (Madokoro family)

Over fifty years after their keels first touched the waters of Coal Harbour, the *Tracey Lee III* and the *Great Northern 5* continue to do the work that they were designed and built to do. "I've got a picture of her on the wall in my house," says Keith Lansdowne, a former owner of the *Great Northern 5*, in recognition of the special place that the product of Kuramoto and Madokoro's careful work has in his family.

Jitsuji Madokoro's drawing for the Thelma S, built for a Prince Rupert fisher and since renamed Tracey Lee III.

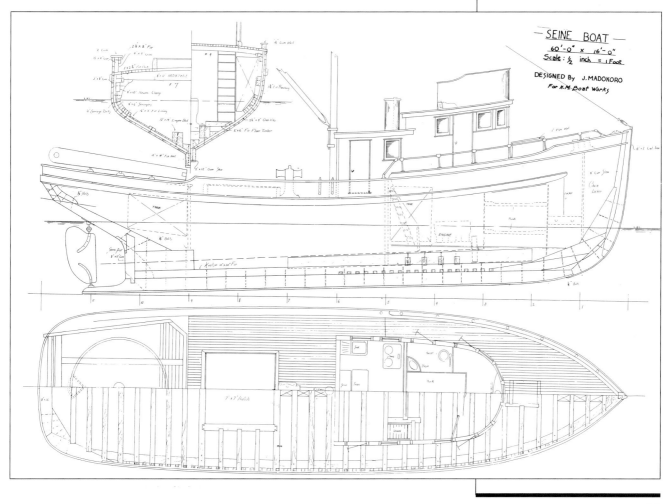

THE SCHOONER *TYEE*

Advances in canning technology in the early years of the century allowed the development of a large-scale salmon fishery on the Pacific Coast. But halibut are not suited to canning, so the early marketing of this fish depended on the development of refrigeration and railways. In 1888 Captain Sol Jacobs brought three New England schooners around to Puget Sound to fish halibut. A number of East Coast fishers came with these boats and brought with them the dory longline fishing methods that had served them well in the eastern cod fishery. This method used skates, each of which is made up of 1/4-inch ground line about 250 fathoms long, rigged with gangen a few feet long with a baited hook on one end. The gangen were attached to the ground line every few feet. Two men in a dory set these skates of gear, left them to "soak" for a time, then hauled them up from the bottom, ideally with a fish on each hook. Sailing schooners like the ones brought around from the East Coast carried about six dories each.

With the arrival of the railways, Vancouver and Prince Rupert became important halibut ports. More boats were brought around from the Atlantic Coast, including some steam-powered vessels like the *Celestial Empire* and the *Flamingo*, and vessels began to be built on the Pacific Coast as well. In total, about 150 wooden-hulled boats of what came to called the "schooner" type were built between the 1880s and about 1930. Most were powered with early gasoline or diesel engines, but they were also rigged with sails which gave the boats a little more speed and stabilized them in the heavy seas that they routinely fished. Trip lengths were limited by the amount of ice that these boats

THE SCHOONER *TYEE*

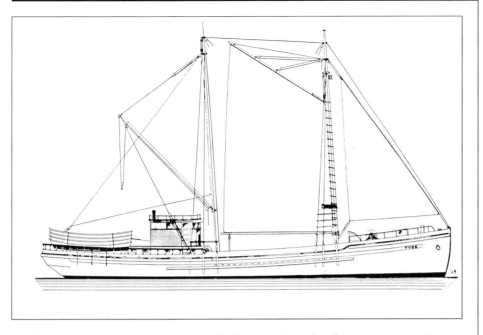

The classic and understated lines of the schooner *Tyee* as they were drawn in *Pacific Motor Boat* magazine in 1914. Note the dorys nestled on the stern.

could carry as it was some years before on-board refrigeration was introduced.

In the period 1910–1914 the Puget Sound yards were especially productive, launching a total of fifty-three schooners. This building frenzy was due in part to the discovery of large concentrations of halibut in the eastern Gulf of Alaska. One of the yards building at this time was Nilson & Kelez in Seattle. Andrew S. Nilson followed the Norwegian tradition of boatbuilding, which combined an understanding of woodworking with the hard-won experience of fishing the North Atlantic. Nilson and Kelez built ten halibut schooners between 1910 and 1914, the largest of which was the 103-foot *Tyee*.

Originally powered by a 150-hp Standard Gas Engine, the *Tyee* was built for Scott Daniel and the Moe Brothers of Poulsbo, Washington, and went fishing for a firm known as Poulsbo Fishing. The owners were loggers; perhaps they thought a halibut boat would be a good investment. Things appear not to have worked out, however, as reports show her suffering serious hull damage when driven onto the beach at Gakutat in 1915. Two years later the boat was sold to the New England Fish Company, parent of the Canadian Fishing Company and a leader in the halibut fishery.

In 1920 she was sold again, this time to C.L. and H.C. Hibbard and Olaf Swenson. The new owners changed her name to *Chukotsk* after a Russian landmark in the Bering Sea where, under her new name, the vessel worked at fur trading. This was a task for which her size and heavy construction suited her well. During this period the boat received her first diesel engine.

The vessel returned to the halibut fleet in 1927 under the ownership of veteran halibut skipper Olaf Hvatum, who renamed her *Dorothy* in honour of his daughter. While Hvatum was primarily a fisherman he did charter work on occasion. In the October 1929 issue of *Pacific Motor Boat* there is an account of a hunting expedition that Captain Hvatum made to the Bering Sea. Leaving Seattle on June 17, he travelled to Dutch Harbor in

The *Tyee*, carrying the name *Dorothy*, at anchor in Mitrofania Bay on the Alaska Peninsula in 1929. Note the furled sail on the main boom. (IPHC)

the Aleutians and then to Nome, where he picked up two hunters who were intending to collect walrus and polar bear specimens for the Field Museum in Chicago. The *Dorothy* broke her rudder in ice and had to return to Nome for repairs, but the hunt was a success nonetheless. The article described the conditions: "The *Dorothy*, the largest vessel engaging in halibut fishing on this coast, is powered with a 270-hp 4-cylinder Bolinder 2-cycle semi-diesel which gives it a running speed of approximately 10 miles per hour. The Bolinder powerplant showed its capabilities when the *Dorothy* had to make its way through the ice floes, driving the sturdy vessel's iron-sheathed stem with smashing strength through the ice as it drove the *Dorothy* northward. Often the vessel would be stopped after a strong drive, and Chief Engineer Bill Knutson would back the ship in its own wake, and then at the highball ahead bell from the pilothouse, 'hook the big diesel on' and with the added momentum, the ship would drive through the wall of ice."

In the early 1930s the *Dorothy* got involved briefly in the cod fishery. Then she was sold to the Northwest Marine Dredging Company of Portland, which used her as a sea dredge for gold-seeking operations at the mouth of the Rogue River in Oregon. By 1940 the *Dorothy* was back where she belonged, fishing cod, this time with Robinson Fisheries of Anacortes. In the early 1950s she returned to the freight business, carrying general cargo between the Columbia River and Alaska for the Alaska States Shipping Co.

In the mid-1950s, when a number of larger American fishing boats, mostly seiners, were brought into Canada, the *Dorothy* was among them. John Kovac brought her up in 1955 and registered her in Canada as the *Shirley Rose*, then sold her two years later to John Cooney, owner of the Ferry Meat Market. During this period she had several skippers, including Captain Hughie McLean, who later went on to become a successful troller.

With the implementation of the 200-mile fishing zone in 1979, Canada lost its place in the Bering Sea and the Gulf of Alaska halibut fisheries and the *Shirley Rose* lost the purpose for which she was built. A series of owners tried to make a living with her in a variety of ways, but in 1981, when she was seized

The *Tyee*, now carrying the name *Shirley Rose*, rigged for longlining and with a monstrous house installed. This was how she looked the first time Brad and Karen Scott bought her. (FT/CI 4525-5)

by the bank and tied to the dock with several thousand dollars owing on her, she looked to most people like one more victim of soaring interest rates. The bankers didn't care, and probably didn't know, that this tired old hulk was once one of the largest schooners in the American halibut fleet. Being "downtown" people, they probably did not understand the potential value of her new engine and limited-entry black cod licence.

At this point Brad Scott arrived on the scene. Brad grew up on Boundary Bay and had his first rowboat by the time he was eight years old. In 1972, when he was fifteen, Brad bought his first commercial boat, a 36-foot ex-fish packer powered by an 8-cylinder Chrysler Royal gas engine. He had to take in a nineteen-year-old partner because he wasn't old enough to get a beachcombing licence. In the next few years he worked on numerous boats. At one point he ran the gillnet collector *Ballerina No. 1* (ex-*David D*), a classic 34-foot double-ender built in New Westminster in 1938. Along the way Brad developed a deep appreciation for the beauty and integrity of wooden West Coast work boats. He bought the *Strady VI*, built in Kelowna in 1942, and he and his wife Karen packed fish in the Gulf and Fraser with her. Cash buying fuelled the entrepreneurial spirit that was growing in the young man.

In 1981 it took a lot of courage to bid on a 103-foot schooner that had suffered many setbacks since she was first launched in Seattle in 1914. There were a lot of fish boats tied up with sheriff's notices on them and Brad had a wide range to choose from, but he put a $25,000 offer on the *Shirley Rose* and got her. The boat had been tied up two years earlier, just as she was about to leave on a black cod fishing trip. It was nice that her fuel tanks were full, but she also had a galley full of grub and eight tons of herring bait in the hold. "Karen and I wrapped diapers around our faces and went into the galley and just scraped the food off the shelves with spatulas," Brad recalls. "When we took the bait out we had to bleach our hands and burn our clothes."

His intentions were primarily mercenary. He sold the black cod licence, then stripped the engine and electronics out of the boat and sold them. The more he worked around the boat, the more he became impressed with the soundness of her massive hull. The beginning of a dream was born. But it didn't mature rapidly enough and Brad sold the hull to a church group which towed it around the coast for a year. They in turn sold it to a fellow who built a square second level onto the cabin and lined the whole thing with drywall. The *Shirley Rose* was on that path all too familiar to the classic boats of the coast. Her days were numbered; it was just a question of whether she would be towed out to sea to be sunk with dignity or end her days rotting on some mud bank.

The second time Brad and Karen Scott bought the *Tyee*, someone had begun making her a live-aboard by installing a pair of sliding glass patio doors. The Scotts had their work cut out for them. (Brad Scott)

Brad, who had gone on to develop a successful fish-buying business in Port Hardy, had not forgotten the boat. In fact, one of those dreams that people who don't understand them might call a disease was growing in him. He could not get that beautiful hull out of his mind, and seeing what was happening to her made it all the worse. Finally, when the drywaller's plans started to sour, Brad got his chance. He bought back the hull of the *Shirley Rose* in 1983 for a little less than he had paid the first time. He had the dream and he bought a mess. She was seventy years old, with no power, rotting decks, a disaster for a cabin. Her huge fo'c'sle had been foamed and glassed as a quick-freeze room for the black cod. Brad planned to make a charter yacht out of the venerable hull. If people doubted his sanity the first time he bought her, they were given real cause to worry now.

The hull was already stripped of engine, wiring, plumbing and rigging, but Brad's first job was to strip it some more. He hired a crew and removed most of the wheelhouse and extra cabin, then stripped the foam and glass out of the fo'c'sle freezer room. "We took three thirty-yard dumpster loads plus numerous pick-up loads off the boat," says Brad. "She came up a foot in the water."

Brian Falconer, who had bought and restored the classic yacht *Maple Leaf*, came to work on the project with fellow shipwright Wayne Martin. The main deck was good, but the forward and aft decks were badly rotted and had to be replaced, along with a number of deck beams. This sealed up the hull, and together with a cocoon of tarps, allowed work to continue through the winter. The bills were climbing but the big job was well underway. Brad believes that one of the reasons the boat survived so well was that she was built of well-seasoned wood, so taking his lead from the original builders he scoured demolition sales for aged lumber. For example, the old Cambie Street Bridge provided fir for deck beams, and he bought timbers from the old Timberland Mill, which had been built the same year as the schooner.

Then came one of the worst days of Brad's life. "I had taken my wife out to dinner in Steveston and we were driving home when I decided to stop off and see the boat where she was tied at the Pacific Coast Camp. When I got to the dock I saw all of the fire trucks and someone told me that the big old boat at the end of the dock was burning up."

Fortunately, the fire was put out before it had gotten a good hold on the boat. The new aft deck and the bulwarks were badly charred and the back was burned off the house. The fire department was suspicious that Brad might have set the fire to get the insurance, until he explained that he had no insurance. Damage was appraised at $70,000. "I was discouraged and wanted to sell the boat off, but then I decided that this was just another hurdle."

The fire did precipitate a change of plans. Brad decided to finish the boat not for chartering but as a fish packer. This was more expensive, as it required a larger auxiliary engine to power the hydraulic winches. It was also necessary to convert the wooden-lined fish hold for refrigeration. To do this an insulating layer of foam was put in and lined with a hard fibreglass overlay. Piping to pump the refrigerated water was also installed. Brad owned an Allis-Chalmers 25000. He lowered this into the fish hold and slid it aft into place in the engine room. It produces 315-hp at 1650 rpm and turns a 68-inch wheel through a new 514c Twin Disc 6:1 gear and about 25 feet of 5-inch shaft. The hydraulics and charging system for the 32-volt power supply run off a 200-hp Isuzu. There is an additional one-cylinder Ruston for charging the batteries. The electrical work was done by John Brown, one of many tradesmen who took the old boat to heart and made her rebirth possible. All new tanks were installed to carry 4500 gallons of fuel aft of the engine room and 1000 gallons of fresh water under the fo'c'sle.

Brad had the hold foamed and glassed for packing fish in slush ice and installed a 3-inch pumping system for flooding and discharging water. This fit the boat for hauling farm fish. An interesting innovation in the hold is a removable T-bulkhead. This divider of heavy aluminum plate creates three compartments for packing fish, but it can be unbolted and removed to allow pallet loads of dry cargo to be lowered into the hold and stored. "You can't spend this kind of money without keeping all of your options open," explained Brad as the boat neared completion. "I'd like to try packing fresh water, farm-fish feed, general freight, or anything that will keep the boat working."

The construction of the 103- by 21-foot hull is massive: The 12-inch frames are made from 5- by 6-inch sawn fir flitches with 3 1/2-inch planking and 4-inch lining. While the hull was remarkably sound it did require some work. Five frames and about forty planks were replaced. About three-quarters of the boat was recaulked and refastened. All new guards were installed. Much of this work was done by Steve Craven at Progressive Shipyards.

The rebuilt main cabin is 29 by 12 feet. The interior finishing is all wood and was done by Gordon and Calvin Wahl of the famous boat-building family. The joinery is of a quality that you would expect on a classic boat. In the aft wall of the galley, two heavy doors have been installed on the built-in freezer and fridge. In addition to the galley and mess area, the main cabin includes a skipper's cabin with double bunk and wash basin. There is also a smaller cabin with a set of bunks. The wheelhouse is raised five feet above the main deck and provides good visibility forward and aft. The old halibut men who sailed her through the first half of the century would be dumbfounded by the full array of modern electronics.

Perhaps the most impressive interior of the boat is the fo'c'sle. On the 1914 version of the *Tyee* the deckhouse was much smaller and the living quarters for the crew of dory fishers, complete with galley, was

The *Tyee* as she looks in the 1990s with Henry Helin's modern Trans-Vac fish pump aboard. As the boat reaches eighty years of age, Henry, who is less than half as old, finds that he can maintain her indefinitely for about $20,000 per year. He uses her to pack both salmon and herring. (Helin family)

located in the fo'c'sle. In the contemporary version this area, measuring 24 feet from the bow post to the bulkhead, has six roomy bunks. It is entirely self-contained with an oil stove, sink, icebox and a mess area for six. A head and shower were installed. From the fo'c'sle deck to the deck head is at least 8 feet. This is fo'c'sle living in the grand style.

The first summer after her rebuild, in 1988, the boat packed salmon for Ocean Fish. Henry Helin took over as skipper toward the end of the summer and had nothing but praise for the boat, which packs 175,000 pounds of salmon in slush ice or about 125 tons of herring. In October of that first season Henry Helin crossed Hecate Strait into the teeth of a full gale, then recrossed with eighteen-foot seas quartering on the stern. On his arrival in Rupert he phoned Brad with unreserved praise for the sea-keeping abilities of the boat.

To build a new steel boat of 70 feet with the same packing capacity in 1988 would have cost about $750,000. The *Tyee No. 1* project was completed for about half that cost. But Brad is a project person, so when Henry Helin fell completely in love with the boat he agreed to sell her to the young skipper. She stands as a testimony to those who built her, those who have sailed her, and those who have cared for her.

A LIFETIME BOAT

Most fishers can recall at a moment's notice the names of all the boats on which they have worked over the years. For some this is a long list; for others the years may be long but the list of boats is short. Ray Michaelson is one of the latter. Born in Prince Rupert in 1920, Ray was only eight years old when the keel of the halibut boat *Melville* first touched the waters of Rupert Harbour. Built at the Prince Rupert Drydock and Shipyard for John Iverson, she is today one of the finest remaining examples of that era's many fine halibut boats. From her plumb bow stem, through her flat sheer line to her rounded stern, her slim 58-foot hull owes much to an earlier generation of sailed halibut schooners. But when built, she was equipped with one of the new winches that took a shive assembly to haul the longline gear over the side so that fishers didn't have to go off in dories to haul the gear. Designed by a Scot who worked for the Prince Rupert Drydock, she has at least one sister ship, the *Covenant*, but similar boats like the *Zapora* and the *White Hope* were built in Rupert while others like the *Velma C* were built and sailed from Vancouver.

In 1934 Ray Michaelson completed the eighth grade and left school to get a job. In Prince Rupert this meant going fishing, which he did on a series of salmon trollers and packers. When he was seventeen years old, and the *Melville* was nine, he was taken on as an in-breaker. This is the halibut fisher's apprenticeship. "In those days it was up to the crew if you got paid—they might give you 5 or 10 dollars each," Ray recalled in 1988. "I made 50 bucks on my first trip. You had to go two trips to 'break-in'. Then the crew had to take you up to the union hall and sign you up. If they didn't figure you were good enough you got cut out."

After successfully in-breaking, Ray had to go and get a job on another boat as there wasn't room for him on the *Melville*. But the following year both Ray and the owner, John Iverson, moved to Vancouver and Ray got a steady job as deckhand on the *Melville*. The halibut fisher's year was really a series of fishing seasons, with halibut starting about mid-March and running through to August. Through the fall and early winter months they longlined cod and dogfish and other species. Then in late winter they made a couple of black cod trips before going back on the halibut. As Ray recalls it, the amount of halibut taken by a boat was determined by the size of its crew. A usual allocation would be 2200 pounds per man per trip in the waters off British Columbia or, if the boat went farther north to Alaskan waters, it was allowed 2800 pounds per man. This encouraged maximum employment and mandated large

The *Melville* sports a fresh coat of white paint from her classic fine bow to her halibut stern, which is designed for weather. A seine boat stern is primarily designed to support a heavy net but a longline boat carries relatively light gear so it can have a finer stern.
(FT/CI)

crews. A far cry from the modern trend to individual quotas for the boat that can now be transferred from boat to boat so that few people can get the most benefit from the boat's capital cost.

With the outbreak of World War Two, Ray went into the air force, but immediately after the war he was back on the *Melville*. A few years later the original owner died and in 1952 Ray bought the boat from his widow. The boat was now twenty-four years old and the man was thirty-two. John Iverson had maintained her in immaculate condition and Ray continued to lavish the same care on her. She had been built with a 75-hp Fairbanks-Morse, which Iverson had replaced in 1940 with a more modern 90-hp version of the same make. In one of the few changes that he made to his new boat, Ray replaced the 90-horse engine with one of the new 6-cylinder Caterpillar D337 engines that were gaining popularity on the coast in those years.

A   LIFETIME   BOAT

From the plumb bow post, to the narrow 14.7-foot beam with its slight taper aft, to the fine stern sweeping up to become nearly round at the gunnels, the hull of the *Melville* is the classic West Coast halibut boat of her era. She was built to longline only, unlike combination halibut-salmon boats with their timbered sterns built wide to carry the heavy salmon seine nets.

A number of these halibut hulls have survived the ravages of time and nature, but few have survived the work of owners who have replaced or extended the original small cabins. The cabin on the *Melville* is the original, containing only the roomy wheelhouse and a companionway to the engine room, through which access can be gained forward to the fo'c'sle. The long narrow fo'c'sle contains bunks and the galley, including a full-size oil stove, brass hot water tank and sink. The table is built into the triangle formed by the bunks. A galley located below-decks, while once the norm on fishing boats of this size, is now the exception as more and more of the old boats have had large deckhouses built to make room for a galley and skipper's cabin on the main deck.

Ray has kept the original mechanical shaft-driven anchor and deck winches. The anchor winch was added in 1940. Before that the anchor had been raised with the American-built Rowe deck winch. This winch has a cleat mounted on a plate on the top. For long-line fishing, this round top-plate could be removed and replaced by an assembly that had a sheave for hauling longline and a smaller high-speed sheave for hauling the sounder lead. Even after the advent of electronic echo sounders, the lead was used to check the makeup of the sea floor. A dab of butter was placed in the cupped bottom of the lead so that a sample of the sea floor was brought up with the lead. Unfortunately, the winch's longlining assembly was stolen from a storage shed a few years ago.

Through the 1950s and 1960s Ray followed the halibut fisher's routine, longlining halibut in season and then longlining black cod and "scrap" fish when the market allowed. He fished tuna most years, trolling a jig through the blue-green waters favoured by those fish. Some years this meant going 200 miles offshore; other times the waters brought the fish within a few miles of the Canadian shore. "My biggest trip was sometime around 1970 when one deckhand and I took 20 tons of tuna in five days within seven miles of Cape St. James," Ray remembers. "That was five days of long arms." During the salmon season Ray packed fish to the canneries for the Anglo–British Columbia Fishing Company. Some years he tied up

Ray Michaelson in the wheelhouse where he has spent a good part of his life. (AHB)

While many older boats have had large cabins added to accommodate the galley at deck level, Ray has kept the *Melville*'s original fo'c'sle galley. (AHB)

Halibut are routinely fished in open waters during harsh weather. (BG)

the *Melville* during the winter months so that he could skipper a seine boat for herring fishing. In the 1970s, with the advent of the 200-mile fishing zone, Canadian boats were excluded from the Gulf of Alaska and the Bering Sea. In Canada only the 400 or so boats that had a recent history of halibut landings qualified for licences in the limited-entry fishery that was introduced. Unfortunately for Ray, when the limited entry was initiated, he had taken some time off from halibut fishing and so didn't get a licence. It is no small irony that today one of the finest halibut boats on the coast is rigged to troll salmon. Ray also packs herring with his boat.

The most remarkable example of good care resulting in longevity on the *Melville* is the mast, which is the original. It has been scraped and revarnished about every three years and the base block has been drilled and flooded with fuel oil from time to time. Most boats of this age have had several masts and now sport a steel or aluminum one. Ray's mast has survived rot because instead of clamping a metal band around it to hold the guy wires, he has retained the older style of spliced eyes held in place with hardwood wedges. This prevents moisture retention in the wood of the mast that is common under a clamp.

The small pilothouse is set about 12 feet back from the bow of the boat. Even with this distance Ray says that storm waves have twice taken out the wheelhouse windows. Once it was a freak wave in a 30-mph westerly. The other time it was a full 70-mph southeaster. But Ray has total confidence in his boat's ability to carry him through anything that the Pacific can blow up. He says that when he gets in a real blow, "as long as we're far enough out to sea I just shut her down. She lays with her stern quarter up the sea, beautiful. But now I'm getting softer and I run in to shelter."

Two years ago the insurance company required that one of the original hull planks be pulled for inspection. Ray says that they had to chisel the plank out, it was so tight. The plank and nails, over fifty years old, were just like new.

The *Melville* was built to fish halibut with a seven-man crew. Now when she trolls for salmon, Ray works her with only one crewman and he can handle her perfectly well by himself. When he takes her from her mooring at Vancouver's False Creek Fisherman's terminal on a low tide, only a few feet remain between her stern and the bank, but he handles her as easily as a pizza delivery man in a Toyota. A man who has lived his life with a single boat has a marvellously easy manner on board.

It was in 1928 that the Model A Ford replaced the Model T on the assembly line and the roads of America. Hailed as mechanical marvels at that time, none could keep up alongside contemporary cars the way the *Melville* does amongst the fleet of modern fibreglass, aluminum and steel fishing boats on Canada's Pacific Coast.

FINAL RESTING PLACE

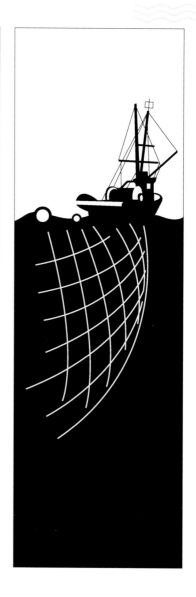

Not many visitors to Steveston's historic waterfront care that the old double-ended gillnetter which sits high and dry over the bar in a seafood restaurant was once a working fish boat. Built during World War Two, and owned for many years by Ken Hagen, the *Vickie* was spared the fate of so many old-style wooden boats which end up disintegrating on a mud bank or rotting in a dryland storage lot. She found her final resting place as a tourist attraction. But the *Vickie* is actually a museum piece: a noteworthy example of the fine boat design brought to the West Coast from Scandinavia.

Ken Hagen was twenty years old in 1926 when he left his home town of Vestness, near Molde on the west coast of Norway, and joined his older brothers Otto and Sigurd who were handlogging in Seymour Inlet. The Hagens had been a farming family back in Norway, but in BC they adapted to the annual round of logging in the winter and fishing in the summer. Ken did both, and began to build a small nest egg. He married and brought his wife Vickie to the Inlet where they moved into a new float house. By 1933 they had amassed enough money to buy a proper gillnet boat.

Ken found just what he wanted in Burnaby, where a builder named Brovold had a nice little 28-footer for sale. The hull showed its origins in the Columbia River-type skiffs of the same length that were used on the Fraser River in the opening decade of the century. But this new boat had a cabin over the forward half of the hull made with vertical

The *Vickie* as she looked on the fishing grounds. (Hagen family)

tongue-and-groove planks. A doghouse sat up at the aft part of the cabin to give space for one person to stand over the engine and out of the rain to steer the boat.

After paying $375 for the hull and cabin, Ken had the boat hauled to the New Westminster waterfront where he installed a Star car engine and launched her as the *Goldie*. He later replaced the original engine with a 7-hp Vivian and fished the boat,

both gillnetting and trolling, until 1949. By that time the coastal forests had been taken from the handloggers and put into corporate-controlled tree farm licences, and Ken had moved down to New Westminster. In 1949 he sold the *Goldie* and went to work with one of his brothers in a store on Front Street. But three years cooped up indoors was enough; he wanted to go back to the inlets and the sockeye. When he heard of a good boat for sale across the river at Annieville he went with Vickie and their two

Vickie and Ken Hagen. (AHB)

children to have a look. The 33-foot double-ender had been built in 1942 by Chris Remmem, who came from Ken's hometown of Vestness. Ken had known the Remmem family as he was growing up; it seemed right to be buying a boat that carried the old-country tradition from the fjords to the inlets. What's more, the original owner was Bernard Larson, whose father had had a hand in founding the Norwegian community at Annieville around the turn of the century.

Ken and his family liked the boat right away. It was longer than the *Goldie*, and a good bit beamier, so there would be more comfort and seaworthiness. While still a double-ender, she had a roomy cockpit for a man to stand in and take fish from the net as it came over the stern to the powered drum. Equipped with a new high-speed Chrysler Ace gas engine, she cost Ken $4500.

For the next thirty-four years Ken fished the *Vickie* out of New Westminster. He used it to earn the money to pay for a house and put the kids through school and get ready for retirement. When it came time to sell the boat in 1986, he faced a problem. Owners of larger boats have a tradition of selling their boats to the engineer as someone who will care for it properly. Owners of smaller vessels hope to find a new owner who will lavish the appropriate respect on a cherished partner. But Ken did not have much hope of finding such a person. The *Vickie* was an old-style wooden double-ender without the ability to pack large loads of fish in an insulated hold. She didn't offer the low maintenance cost of the modern

The *Vickie*, waiting for renovations to begin. (AHB)

wood and aluminum boats, nor was she fast enough to speed up and down the coast to the short openings which had come to be the industry norm. The *Vickie* did come with a valuable salmon A-licence which would go nicely on one of the new, high-speed bow-picker gillnet boats being pumped out of the molds in a number of shops. In fact, the licence was worth more than the boat and could be sold easily, leaving an unlicensed boat which could

The new owners cut the stern away and removed the engine. (AHB)

not go fishing unless someone bought a new licence for it. More often than not, such boats are scrapped.

What saved the *Vickie* was the gentrification of Steveston. The public loves the romance of the fisherman's wharf. This often runs counter to the best interests of the commercial fishing community, which is crowded out of the moorage by mushrooming condominium developments, restaurants and gift shops. The province's oldest fishing community at Steveston has tried to accommodate the competing interests. Working boats moor in front of restaurants designed to reflect the salmon fishing heritage.

With her engine removed and her stern cut away, the *Vickie* is "moored" over the bar in such a way that customers can enter the boat from a mezzanine floor and sit at a table in the same hull that carried Ken out to the fishing grounds for so many years. When told that busloads of tourists were coming to eat fish and chips in his old boat, Ken only laughed. "I don't care what they do with it as long as they put it inside out of the rain."

SCOTTY NEISH, UNION MAN

Elgin "Scotty" Neish grew up in the Roaring Twenties and reached maturity in the Dirty Thirties. He is a big man, well over six feet, with a physique and great booming voice and laugh to match. His physical size and nimble intellect combined with his life experiences to give the fishing industry a man feared by corporate managers and followed by fishers. The economic good times of the 1920s helped build a lot of fishing boats, but the companies maintained their control over the fishing fleets by encouraging a combination of indebtedness and racial paranoia. When the economic downturn of the 1930s brought these tensions into the open, Scotty emerged as a voice for the working fisher. But his training had started much earlier.

"We moved to Deadman's Island in the early twenties. My oldest brother, James, got drowned fishing crabs off Point Grey. We used to cook them, then take them up to the open-air curb market at the foot of Main Street. When my brother drowned, my old man had a nervous breakdown and went off to Australia. We went back East about 1923 because he was afraid us kids would get drowned. Until we learned how to swim, then we came back out in 1929. My dad was a machinist, but he never worked his trade out here. My mother cleaned the top two floors of the Dominion Bank Building across from Victory Square. When she went to work she would row across the harbour. After work she would catch the last street-car home. It used to run down to the Stanley Park loop. Number 13, I think it was. Quite often it was a little foggy and it was my old man's job to go down and blow the foghorn to bring her back home. I remember my mother used to get so mad. It was real quiet of course, and my dad could hear her oarlocks and he kept blowing the horn this way and that. And my mother would hear it loud and soft and she didn't know if she was going to it or away.

"Mother used to row us across the harbour to go to school at Lord Roberts in the West End. Old Man Shannon used to take his boat, the *Zero*, out dragging off Point Grey at exactly quarter to nine every morning and we knew that if we weren't into the wharf at the foot of Bidwell Street by then we were going to be late for school. That was in 1923, before we left.

"We used to drag seine for spring salmon in what they called the Deep Hole right off the mouth of the stream that used to come out of the duck pond [in Stanley Park] right behind what is the Rowing Club now. It took a lot of line to get the seine out there. Then we drag seined out on the point of Deadman's Island. There were Yugoslavs living on the island and they used to have little purse seines and fish smelt right up to Port

Moody and across from Brockton Point on the North Vancouver side near what we used to call the Creosote Works. Old Man Ferrario had a float running out with net racks where he used to repair his seines, and "Spanish" Pete Ambrose had his net house there too.

"The first fishing experience that I had was with Joe Oliver. I went with him for dog salmon in the fall of 1929. We went into Eagle Harbour and took a gillnet across the creek, illegally of course, and my share was all the fish I could put into the big skiff. Our net went dry when the tide went out. I just rowed the skiff up the creek with the fish pugh and pughed the bloody skiff full of fish. Then he picked the net up and we went in in the morning. It was dead calm, foggy weather. They were real black dogs, I think we got either five cents or two and a half.

"In the winter of 1929 I went fishing dogfish out of Deep Bay on Vancouver Island. The whole of Baynes Sound froze right across that year and the fleet was froze in. The old *Michael Taylor*, *Amalac*, *Luac* and *Ribac* were hauling the dogfish out of Deep Bay in scows and they used to bring the barrels of bait up. We were getting $7.50 a ton for them. I remember thinking Jesus Christ, if this is fishing, I don't want anything to do with it. We were second highest boat that year and I think we had $129 by the time we paid our groceries and whatnot.

"In the summer of 1930 I came ashore and worked in the cannery at the foot of Bidwell. That's when they started the first Co-op. It was Millerd's cannery. I worked there all that summer unloading. I was so bloody big and tall that as the tide went down they

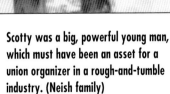

Scotty was a big, powerful young man, which must have been an asset for a union organizer in a rough-and-tumble industry. (Neish family)

would 'Put Neish down there' to throw the fish up to the dock. Martinolich was there with the *East Point* and they had the *Daisy B*. Jimmy Martin was there with the *Curlew M*, I went on her as a deckhand picking up fish from the hand trollers in the gulf. We used to leave every second day for Lasqueti Camp, False Bay, then run up to Campbell River. Leave the next morning and be into Lasqueti that night. Pick up again in the morning, then back to the co-op cannery, unload and back up.

"In 1932 I went fishing with Spanish Pete on the *Balares* with a beach drag seine. We fished for rock cod and perch right up as far as Port Neville, Granite Bay, Buccaneer Bay, Halfmoon Bay, Pender Harbour and Harwood Island. Then he put a beam trawl on that was actually a miniature drag seine with bridles on and gillnet lead line wrapped around to keep the bridle upright. We cut two big trees and spliced them together to make a pole to spread the net. Every once in a while we would hit a snag and break the pole. Then I fished with him with illegal tremel nets with

On the rocks in 1933 with Victor Ferrario's *Lake Como IV*. In the early days many of the inlets were marked on the charts with dotted lines to indicate that they weren't fully surveyed. Groundings were common and the boats usually just floated off. "We did split a plank on the bilge of the *Lake Como IV* once," Scotty says. "Victor just beached her, took a can of lard, smeared it over the split plank, covered it with a grease-covered coal sack, took a floorboard out of the skiff to nail over it all and we finished the week fishing the mainland area." (Scotty Neish)

two layers of 4-inch mesh and 14-inch mesh in the middle. The fish run into it and create a pocket so you get all sizes of fish in the same net. There was us and Jack Troop, he lived on Deadman's Island. The Fisheries officer was Menchions [a relative of the shipyard people], and his ambition was to catch one of us but he never found out that we hid our nets on Trail Island off Sechelt every time we went to town. We used to sell the ling cod to Van Shell Fish at Campbell Avenue, which had just started.

"Then in the summertime I used to take my gumboots across the dock and go seining with Old Man Ferrario on the *Lake Como IV*. It had a 30-hp Vivian gas engine. It was the first year they let the seines fish the Fraser. He had two seines, a shallow short seine for fishing the Nimpkish River. It was only about 170 fathom long and two and a half strips deep. And he had a bigger seine for the straits. It was still open on the Nimpkish but it had fallen off so we were going down to the Fraser River. Going down through Johnstone Straits, be goddamned if we didn't get some real good showing. A couple of boats got some good sets. We had one seine on

SCOTTY NEISH

the table and the other part in the hold and part on deck. Old Man Ferrario decided that he was going to fish this last-minute run of sockeye, so we went into one of them little bays, down just below what they call Fine Beach, to put the other seine in the woods. In the process, with the mast, we broke the bloody telegraph line that ran on the trees along the beach. We never even seen it. Old Man Ferrario he didn't give a damn about that, he just wanted to get the seine off. So we put the boat into the beach alongside a bluff there and manhandled the goddamn seine out of the hold. By that time the tide had changed. We fished there the next day but the run was gone and that was it. So the next day, I think it was a Friday, we went and hauled the bloody seine out of the woods again and down we go to the Fraser River.

"When we got down to the Fraser the fish were up on the flats off the mouth. Ferrario put his deep net ashore at the Imperial Cannery in Steveston and went chasing these humpback up on the flat with the shallow seine from Nimpkish. So we set the net and Jesus did we ever roll up with Dungeness crab. The net was full of them. What a bloody mess, but we got humpies in amongst them so that was fine. Then we ran hard aground with the net out. What a bloody mess. We had to go and put the net ashore and pick all the crab out. That Old Man Ferrario, he was a real wild man when it come to fishing. But he really caught fish. I just seined with him that one year in 1933, then he retired and Louie Benedet took over the *Lake Como IV* and I went out with him.

"Louie Benedet was a harder man. He made fourteen sets one day and that was a record. The days weren't quite as long. It used to open on May 15 for sockeye on the Nimpkish. He was a driver but a good man to work for. He later built the *Mermaid I* and then the *Belina*.

"Then in 1935 I went out with Clato Ferrario. That was the year he bought the *Shuchona No. 4*. Then it was 1936 that Mathews, manager for the Anglo–British Columbia Packing cannery at Glendale Cove in Knight Inlet, offered me the job as skipper on the *La Paloma*. That was the year of the strike and I was quite active in the union [the Fishermen and Cannery Workers Industrial Union]. The strike was in the first of the year [early July] before the humpies started to run. We were trying to get a signed agreement for humps and sockeye. That was part of the problem, we went on strike when there were no fish. It was poor strategy on our part but you learn by

This 1945 photo shows fishers on a weekend working on their nets. Weekends were also the time when union organization best took place without drawing too much attention from the bosses. (NFB/PAC PA-145359)

trial and error. Also, in 1936 the company managers picked on the weak sisters and the skippers of company boats. They know how to pick 'em. They get them in the beer parlours and talk to them. The only white boat that broke ranks and went out and scabbed was the *Optimist*, but the native Indians they broke and then one or two of the white skippers went and the thing fell apart fairly rapidly, although we were on strike for two or three weeks.

"The fleet in those days used to take practically all their own canned goods, spaghetti, flour and sugar. Christ, if you went aboard the boat you could hardly get into your bunk for goods. The skylight was hanging up with salamis and smoked hams. The stoves were all coal burners with stacks of coal up on the bow. So when it come to the company stores shutting off credit, it was mainly fresh stuff that we were short of. We were all tied up in Alert Bay, but the Sointula Co-op Store offered us credit over there so we used the *La Paloma* to go over and pick up the groceries for the whole fleet. The women in Sointula were cooking fresh bread and buns for the whole fleet.

"It's hard to organize seine boats. You're on a share agreement and the company dominates. If you were too strong a union man, you had trouble getting jobs. Like the year I fished with Clato I was organizing for the union and the next year he wasn't going to have me on the boat because I was a troublemaker. In fact, in 1938 when I was skipper on the *Chief Takush* I came back aboard the *Shuchona* to do organizational work and Clato ordered me off the boat. Louie Benedet always took a good position though. As a vessel owner he relied mainly on the price of fish for his money and the union fought for increased prices so it was in his interest in the long run to get good prices and he wasn't anti-union at all.

"In 1938 we had the strike in the falltime. It was just about the last of the sockeye season. It was a case of negotiating for 10 cents apiece for fall chums and 8 cents a pound for coho. We learned from the experience of 1936 and one of the first things the strike committee done was move the whole fleet offshore and anchor them bow and stern in two lines right

The seine fleet tied up at Granite Bay on the top end of Quadra Island during the 1938 strike. (Scotty Neish)

across the bay at Alert Bay. Luckily there was no real bad weather. You couldn't go ashore without the committee's permission. We kept everybody under pretty close scrutiny so the company managers couldn't get at them.

"Of course, the key thing was the native Indians stayed with us, that was the first time we had solidarity. In Alert Bay they stayed tied at the dock but pledged their support. When we came down to Granite Bay on the north part of Quadra Island, Dan Assu [Chief Billy Assu's oldest son] came over and pledged that they were going to stay with us through the whole thing. The Native fleet was tied up at Cape Mudge and Quathiaski and that more or less saved the day.

"The other thing was that the companies forced us to put the nets ashore. We had said that we were determined to call the season off if we had to. The companies said, 'OK, you've got to put the nets ashore.' So the strike committee said to strip the nets in order to prevent the companies putting them on other boats with scabs.

"You see, when nets were made of cotton web and manila lines, we had to take them all apart for storage at the end of the fishing season. We cut the lead line, the cork line and all the strips of the seine and put the nets ashore in pieces. The company tried to stop us. We went ahead and stripped the seines anyway.

"Then we went to town [Vancouver] from there. On the way to town, the boats that had to deliver the nets to Bones Bay, we arranged for them to take them up, then we would meet them in Port Harvey, just below the mouth of Havannah Channel. Then we went down to Granite Bay and into Quathiaski and had the meeting with Dan Assu. Then we went down to Blubber Bay and that's where the IWA was on strike. That was a breakthrough for them, and we went in to show solidarity with them. It wasn't by design but as it happened the Union steamboat *Chelosin* was coming up with a load of scabs for

The *Chief Takush*, with Scotty in command, led eighty vessels of the striking seine fleet into Vancouver Harbour in 1938. (FT/CI 699-5)

that night. The *PML 14*, it's the packer *Texada* now, was in there with the provincial police. When the *Chelosin* came in we filled the bloody dock with fishermen—it was standing room only—and she turned around and went back and we heard they unloaded the scabs in Powell River.

"The next day, just at daylight, we steamed down the coast. It was dead calm, flat as piss on a plate. We all come together right off the lighthouse at Point Grey. There was about eighty boats by that time. We all agreed we would steam into Vancouver in line-ahead. I was running the BC Packers boat *Chief Takush* that year and I went first. Then there was Mike Sakich on the Nelson Brothers' *Gradac*, he was the second boat. After that the companies settled with the union. The fleet rehung the seines and returned to the grounds. I took the *Chief Takush* out to Barkley Sound and fished out of the Kildonan plant, then went down to Nitinat Lake for the dog salmon run there.

"That was the first successful seine strike because we had solidarity with the natives. But when I come back to get the boat in 1939, McKenzie, the BC Packers manager at Alert Bay, wouldn't give me the *Chief Takush* back. He wanted me to take the old *Wadsulis*. It was a real haywire bloody boat. Tommy Hunt used to run her and, of course, he wouldn't get my boat because he was Indian and Indians got all the junk in the bloody fleet.

"When we signed the agreement in 1938, one of the clauses was that there be no discrimination against anybody for union activity during the strike. Of course, I was in the leadership. I was secretary of the local at that time. We had a Canadian Labour Congress charter. I told McKenzie right then that I thought it was discrimination from the year before.

"'Oh, no!' he said, 'It's just that you didn't catch enough fish last year.'

"I said, 'I want to see somebody else.'

"'No,' he says, 'you've gone as far as you can go. That's it. You're finished. You either take the *Wadsulis* or else.'

"'OK,' I said, 'yeah, that's fine.' So I turned around and walked out of his office. As soon as he went away I walked back into the office and asked to see Buchanan, he was the president, about personal business.

"He called me in and I told him about the discrimination and he called McKenzie in. Well, Jesus Christ, when McKenzie seen me in Buchanan's office he hardly could control himself.

"Buchanan said, 'What's this about you not giving Neish back the *Chief Takush*?'

"McKenzie said, 'Yeah that's right, he's got to take the *Wadsulis* out.'

"'Well, how was his production?' he asked. McKenzie had to admit that I wasn't no highliner but it was fair production. So Buchanan said, 'You better give him his boat back.'

"So I got my boat back, but I couldn't get a goddamn thing out of the company net loft, a tow line or spare piece of web or a damn thing. I wasn't a very popular guy. But I had that boat right up until the war was declared.

"I didn't go back seining after the war. I figured if you're a company skipper you can just go so far. You can't be very independent unless you were a highline producer. I wasn't a hard enough driver to be a highline production man like Frank Ferrario, Louie Benedet, Anzulovich and all those guys. They can call the shot.

"So I decided after the war that I wasn't going to go back into the seining and I bought a troller and went fishing with my brothers."

Scotty spent much of his life speaking out for the workers of the BC fishing industry. (UBC)

THE GUMBOOT NAVY

At the Battle of Zeebrugge on the coast of Belgium during World War One, the vessels of the commercial fishing fleet played a major role in clearing the area of mines. In related actions just down the coast at Ostend, a young Canadian won a Distinguished Service Order and a few days later, a Victoria Cross for valour in action. Lieutenant-Commander Roland Burke had tried to enlist in each of the Canadian forces at the war's outbreak, but had been rejected because of his poor eyesight. Desperate to do his bit, the young man returned to Great Britain where he had been born and where he was successful in joining the Royal Naval Reserve. He was put in command of a small motorboat which was part of a force raiding the port of Ostend. On the second of the two actions, this one on the night of May 10, 1918, Burke took his boat into the harbour to rescue survivors of the attack. Working under heavy machine gun fire, he pulled three British seamen from certain death in the waters of the port. His little wooden vessel, her civilian name obliterated by a military designation, had fifty-five bullet holes, including one from a six-inch shell, but she had engaged the enemy, done her rescue work and come home.

According to Scotty Neish, it was the success of these naval reserve vessels in World War One that prompted the Canadian Navy to consider forming a Fishermen's Reserve unit to patrol Canada's Pacific Coast during World War Two. And it was Lieutenant-Commander Roland Burke, of the weak eyesight, who was chosen to organize it. Often denigrated as a comic fleet of funny little fish boats, the force was designed with serious intent and organized by a commander who understood the strength of boats designed to meet the challenges of local weather and geography.

When Canada declared war on Germany on September 10, 1939, the nation's warships were concentrated in the Atlantic. War with Japan was still over two years away, but some danger existed to the Pacific Coast and it was this threat the Fishermen's Reserve was created to meet.

Recruitment had actually begun in 1938, before war was declared. Scotty Neish recalls the cruiser *Skidegate*, built on a seine boat-style hull but with a full cabin aft, cruising the coast signing up volunteers. The first group of fishers began training early in 1939 and Neish himself was called from the grounds to go into the Reserve on September 6, 1939, four days before war was declared. Each of the major fishing companies was asked to contribute two vessels to the Reserve, and fishers served as crew.

The Fishermen's Naval Reserve quickly became known informally as the Gumboot Navy. Scorned by the regular navy, the members of the

The Gumboot Navy sent fishers out on patrol off the West Coast without a lot of idea what they were to do if they did see a submarine. Sometimes the weather kept them busy just keeping afloat. Scotty Neish took this photo on patrol in 1940. (Scotty Neish)

Gumboot Navy took great pride in the fact that they were not required to salute officers and could leave their fleet of converted fish boats without waiting for official liberty. But most important, they held pride of place for their extensive knowledge of the coast.

Curly Auchterlonie was a rugged skipper whose name crops up in stories of rumrunning in the 1930s, saving sinking packers for the fishing companies, and skippering boats in the Gumboot Navy. Scotty Neish recalls the Scots-Canadian as a "congenial rogue and all round good guy." When the Navy needed to get a bunch of heavy chain up the coast to Prince Rupert, they planned to load it on a barge and have one of the boats tow it up. Curly said, "Hell, we carry more weight of herring than that in the boat. Just load the chain into the hold." The chain was put into the hold by crane and the boat settled lower and lower in the water, until the chain was all loaded and there was an inch of water covering the deck. "Hell," said Curly, "that's nothing. It's just that on the herring we have a deckload of fish so you can't see the water on deck."

Curly left for Prince Rupert under the scornful eye of the regular navy types, who knew all about load lines and the theory of seamanship but had no practical experience earning a living from the sea. Curly re-

turned some time later with the chain safely delivered. Asked how the trip had gone, he commented only that it had been "no problem, I never even took a chart down from the rack." Not only did he know the loading capabilities of his boat, he knew his coast well enough to navigate its entire length without reference to a chart.

Charlie Clarke described Curly's experience running one of the big seine boat-type vessels that the Navy had built for the Gumboot Navy later in the war years. "The boat had one of the new Vivian diesels with a supercharger on it," Charlie recounted. "That was the first time they had superchargers. I don't think they did much good because it took as much power to run the supercharger as you got out of it and they made a heck of a noise when they started up. Curly was on patrol up on the West Coast. He thought he would go way offshore to look around. Pretty soon he saw two boats coming like mad at him. They were coming as though they were going to block him in on the bow. He figured they must be enemy ships so he hollered at the engineer to put on the supercharger to give him more speed. He was going to make a bid for shore. The supercharger started with a hell of racket. Curly started swearing, 'The dirty bastards, they wouldn't even give a man a chance!' He thought they were shooting at him. It turned out to be two American destroyers wanting to know what he was doing out there. 'After that,' Curly said, 'I followed the kelp [close to shore] all the way up the coast!'"

On December 7, 1941, headlines around the world screamed, *Japan Bombs Pearl Harbor*. The attack electrified the West Coast. Suddenly British Columbia, this quiet backwater to world events, became a front line in a new war. If it was possible to bomb Hawaii and land troops on the Aleutians, then the Pacific Coast of North America was threatened. The Gumboot Navy was more important than ever.

This was not unreasonable thinking. But the actions of the Canadian government against its own nationals, Canadians of Japanese descent, was clearly without reason. Both the Army and the RCMP are on record as saying they saw no need to evacuate Japanese Canadians from the coast in 1942. In spite of this, the evacuation proceeded under the provisions of Order in Council PC 1665, dated March 4, 1942. All property was turned over in trust to the federal government and the evacuation was

When the boats were taken into the Naval Reserve they still had their civilian colours—like the BC Packers' *Mina H,* shown here with a Canadian Fishing Company boat. (PABC 64153)

When the *Kuroshio* was taken from her owner Charlie Nakamura in 1942, her Japanese name was changed to *Surf* and the designation FY 24 was painted on her bow. (VPL 16196; VMM)

handled by the British Columbia Security Commission appointed for that purpose. The three-member commission was composed of a representative of the RCMP, the Provincial Police and an industrial representative, Austin C. Taylor, who was also a director of BC Packers.

Among other property seized were 1337 fishing boats, including 141 cod boats, 148 packers, 860 gillnetters, 120 trollers and 68 seiners. Some of these seiners were sold directly to the major fishing companies and their non-Japanese fishermen, but twenty were added to the fleet of the Gumboot Navy.

When these policies were enacted, Charlie Nakamura was a particularly successful fish processor. He had plants for the salting of dog salmon at a number of locations, probably including involvement in the saltery at Blind Channel for which the Cape Mudge people fished in the 1920s. In 1941 he had recently expanded his operation from Telegraph Cove, where he had been salting dog salmon, to Bull Harbour on Hope Island. This expansion enabled him to purchase troll-caught salmon, which he packed to Seattle for resale. He had had two large seine boat-type vessels built at Union Boat-works in 1939 for this purpose. The *Kuroshio* and the *Arashio* were among the first boats with a top wheelhouse in what came to be known as a double-decker style.

"When they grabbed Nakamura and put him behind the mountains, Blain Myers [his bookkeeper] took over and they took the *Kuroshio* and the *Arashio* into the Gumboot Navy," Scotty Neish remembers. "They were named after two of the biggest flagships of the Japanese navy and of course they couldn't have the HMCS *Kuroshio* and the HMCS *Arashio,* so they called them the *Billow* and the *Surf.*

"My youngest brother Angus was on the *Surf* when they ran ashore in a heavy fog on Cape Cook. They were in a cave there for nearly two weeks in mid-winter. There was a big swell running in and they were travelling at night when they ran ashore at full speed and they couldn't get her out. She went over a reef and ran right up in a gulley. Every time they backed up they just hit the reef. Then the swell lifted her and turned her around. They tore the bottom out of her and she sunk right there, but they got all the groceries and everything off. It was the same winter that the *Northholm* went up and only a couple of the crew survived. Anyway, they got in the cave. There are mountains of driftwood there and they must have burned a thousand cords of wood, but the boats just kept going by. No one saw them until the *Bruce I* was fishing black cod in the wintertime and he saw them and came in and picked them up.

"Jimmy Detweiler on his *San Thomas* and one other Fishermen's Reserve boat [the *Leelo*] went back up there and filled her with oil drums. They got her off and towed her down to Esquimalt and pulled her up on the marine ways. She had bad holes in the bottom of her, but she was pretty sound. The navy declared her a total loss, so the fishermen around there bid on her because she could be fixed up. We figure the Navy decided that after they had called her a total loss they couldn't sell her because that would make a fool of them. This is hard to believe, it was before the days of the chain saw, but they actually cut that boat up, piece by piece, and carted it away and burned it. It took the chippy chaps months to saw that bloody boat up. Those fishermen were so bloody mad. The Vivian engine ended up in the diesel training school.

"Spence Turner, he was in the Fishermen's Reserve, he had the

One of the Naval Reserve's early duties was to assist in the seizure of Japanese Canadian boats. Note the rope fender on the front of the *Moresby No. 2*. It was to prevent damage when she tied bow to bow with another seiner in a twin herring fishery. The *May S* was built by Kishi in Steveston for Mr. Saimoto, who named her for his Canadian-born daughter. This boat seined salmon in Johnstone Strait in 1993. (UBC)

Mitchell Bay. He and Edgar Arnet with the *Cape Beale*, all they done was every morning they did a sweep out around William Head for mines. But Spence Turner put his boat up on the marine ways to get it caulked and copper painted. Well, he had what they call a scribed water line where they cut a little groove into the planks along the water line. It makes it easy to always paint a nice straight line. So this bloody chippy chap, this greenhorn caulker, he caulked about sixteen feet of the water line, across three seams. When we came down, there were two guys driving the caulking into this goddamn waterline. Spence Turner was so bloody mad! But that's how green they were."

Turner wasn't the only fisherman to have his boat abused by greenhorn naval types during the war. Charlie Clarke sold his *Western Challenger* to the co-op for use as a packer and arranged to build a new boat under a federal subsidy plan. The government apparently did not let the Indian fishers in on the scheme, but Charlie Clarke was head skipper at Nelson Brothers and Ritchie Nelson let him know that the company would be building a couple of subsidized boats. Mercers Star Shipyard in New Westminster built a number of vessels under the program, including Canfisco's *Cape Russell* (now the *Abound*) and *Cape Mark*, BC Packers' 75-foot *South Isle* and *Eastisle* and Nelson Brothers' 73-foot *Western Mariner* and the *Western Commander*. Charlie had the *Western Girl* built at Menchions Shipyards in Vancouver. Menchions also built the 73-foot *Vic Isle* for BC Packers under the program.

As Charlie Clarke told the story: "Now they had told us that they wouldn't take these boats away from us if we built them. Before my boat was finished Menchions told me, 'There's been a navy man down here looking at the boat.'

"The next time I came down there was an army man looking at the boat. Then the air force was down looking at the boat. So I got my boat launched, through her trials and I went out for two months on the herring. I came back in and they came and grabbed her. We got 20 percent off the price from the government for building these boats if we could get the engine and lumber. They took that boat and I said, 'What about my charter now?'

"Well, that boat had cost me about $32,000. Actually it cost close to $40,000 with the 20 percent. They figured out the charter at so much a thousand and I got 20 dollars a day for my boat while it was in the navy. An 80-foot boat for 20 dollars a day! You couldn't get a rowboat for that today. They had her for two years, then they were going to give the boat back in June. I told them I wouldn't have time to get her ready for the fishing season because we start in June. But I still didn't get her back until June.

"They allowed me $10,000 to put her back in shape. I was out fishing for about three weeks when all the shaft bearings started to heat up. My engineer went back in the fish hold and cracked the bearings on the intermediate shaft and the tail shaft and the bloody thing flew up. So I said to the navy people, 'What the bloody hell has happened here?'

"Well, when they got the boat they had put six ton of metal punchings out of the shipyard into the stern. You see, a seine boat is always down by the bow, so that the seine net can level it off. They put that weight down in the stern so that she'd be level all the time. Being a new boat with a big engine in the bow she just bent. They had her for two years that way. They had raised the engine, cut the pipes off and put shims

The *Surf*, designated FY 24 in the Fishermen's Reserve, aground at Cape Cook. (VMM)

in. So I had to put the engine back on the base. Well, then I had to lengthen every goddamn thing. It cost me over $14,000 to do this. Then I got her up on the ways and sixteen planks are worm-eaten. They'd never copper painted her in two years! I got after them and they said, 'We didn't have time.'

"My lawyer took it to court. It had cost me $22,000 to put the boat back in shape. I had the bills and everything. But the court—no, they couldn't do it. The $10,000 was all I was going to get. But there were no court charges so they knew damn well they were wrong. If I had won the case they would have had to pay to put every fishing boat back in shape."

There were more layers of power and influence. The Japanese Canadians lost their boats, the First Nations people didn't get word of the subsidy, a Euro-Canadian fisherman wasn't paid the full cost of his vessel's repairs.

The extent to which the Navy recognized the value of the seine boat design was borne out when the federal government actually built six seine boats for use as patrol boats in 1941. Built in three different yards, they were given native animal names. They included the *Moolock, Nenamook, Ekoli, Leelo, Talapus* and *Kuitan*. Scotty Neish had command of the *Nenamook* when word came that a fellow who worked on a fish scow in an upcoast inlet had sighted a submarine. One of the grey-painted seine boats of the Gumboot Navy was dispatched to investigate but was unable to detect anything. After several patrols it was decided that the submarine was staying hidden because of the obvious naval appearance of the patrol boats. "Perhaps if it was just an ordinary fish boat the sub would show itself," someone suggested. The *Nenamook*, which had never been a real seine boat, was given a coat of green hull paint and a white house in the company colours of Nelson Brothers. It was renamed the *Western Kid* in keeping with the company tradition of Western as a prefix to boat names. Scotty then went prowling the suspicious inlet. After several days of cruising, the search was called off, but not before several legitimate fishers had been given cause to ponder the strange behaviour of this latest addition to Nelson Brothers' fleet.

When the Canadian government decided to build additional patrol boats during the war, they chose a seine boat design. The *Nenamook* is shown here on the Fraser River following her launch at Mercers Star Shipyard in New Westminster. (PABC 89416)

Fish boats masquerading as navy boats, navy boats masquerading as fish boats—as always, the first casualty of war is truth. The Gumboot Navy was disbanded in May 1944, as the threat of Japanese invasion eased. Most of the men returned to fishing.

FROM NORWAY THE HARD WAY

In 1986 a young Norwegian couple invested all of their savings and a good bit of borrowed money in a steel vessel specially designed to transport farmed salmon. By the time they christened the 120-foot *Boe Junior* at the shipyard in Mjosundet, Norway, the market for this type of boat was already overloaded. Someone told them that fish farming was booming out on the west coast of Canada. Without further ado the young skipper-owner fuelled up his boat with the last of his cash, kissed his wife goodbye and set sail for the New World.

When he eventually arrived in Vancouver he was met by customs officials who told him that there was a 25 percent duty on vessels brought into Canada. He had neither the money for the duty nor cash to pay docking costs. His wife flew out to meet him as he frantically sought financing. But none was to be had and he eventually was forced to sell the boat at a loss to BC Packers, who renamed it the *Western Thunderbird* and put it to work carrying their farmed salmon.

The story of the *Boe Junior*, and the sad circumstances which greeted its owner, bring to mind the story of another, considerably smaller fish boat which made the voyage under very different circumstances. The *Kaare II* was built in Hardanger, Norway in 1917. She suffered a fire some time in the 1920s and was rebuilt to an overall length of 72 feet. By 1940 she was skippered by her owner Ottar Novik, who fished her as a seiner using two powered dories for setting and retrieving the net. A respected skipper with forty-two years' experience, Ottar was particularly noted for his skills in coastal navigation.

Haakon Novik was thirty-one years old at the time and worked as engineer, tending to the *Kaare II*'s 40-hp Wickmann 1-cylinder semi-diesel. In April 1940, Ottar took his boat 200 miles north from his home port of Bremsnes, Nordmöre, for the spring herring fishery. When he heard that the Nazis had invaded Norway he and his sixteen-man crew headed for home.

Shortly after they got there, they were called to help with the evacuation of the nearby city of Kristiansund which was being bombed by German aircraft. "We kept running between Kristiansund and various nearby harbours," Haakon said later, "doing our utmost to move as many people as possible away from the hellfire which once had been the city. We ran every night, but tied up each morning before the planes came back." After several days the *Kaare II* was ordered south to Gjemnes, in the Batnfjord, to evacuate government officials north to Tromsö.

The *Kaare II*'s next job was to haul freight for the military. But

FROM NORWAY

Haakon and Ottar Novik were becoming increasingly concerned for the safety of their families. Finally they persuaded the military to let them go south for their families, although they were allowed to go no farther than the island of Vega. Here Ottar left the boat in the care of Haakon and his brother-in-law, Peder Engvik, while he caught a ride another two hundred miles home. Once there Ottar launched the two purse-seine dories he had stored earlier and loaded them with provisions for the trip north. When the two dories left, they carried Ottar Novik, his wife and six children, brothers Alfred and Ove, Haakon's wife, Karen, and her brother Ingvart.

On the way north they stopped at Titran on the Island of Fröya, where they picked up Ottar's father, who was in his seventies. While they were on the island the Germans overtook them, but the family escaped undetected. They continued northward in the open boats to Vega where the *Kaare II* awaited them. The seine dories were hoisted to the davits and the trip to Tromsö completed without incident.

Back in Tromsö the *Kaare II* returned to work supporting Norway's faltering war effort. In June 1940, when Norway surrendered to the Germans, the family decided quickly to leave the country. They took on fifty barrels of fuel and three barrels of lube oil. The Wickmann engine burned one barrel every fifteen hours, but figuring even two barrels every twenty-four hours they had enough for Ottar's intended destination—Canada.

After beaching and abandoning the two dories, the *Kaare II* set sail along with two other vessels, a coastal freighter and a fishing cutter. On

Above, Haakon Novik in 1987. (AHB) Below, the *Kaare II* as she looked on arrival in Vancouver. (Novik family photo)

The Novik family attracted the attention of the newspapers when they reached the east coast. This old newspaper clipping shows them grouped on the stern of the *Kaare II*. (Novik family)

board the *Kaare II* were twenty-three people: Mr. Novik Sr., his six sons and two daughters, eight grandchildren, four daughters-in-law, and the brothers Peder and Ingvart Engvik who were relatives by marriage. The other two vessels left a little ahead of the *Kaare II* but travelled slowly until all three boats were together. "All of us were headed for the Faeroe Islands," wrote Ottar Novik in Norwegian years later, "but we didn't head in that direction at the start because we wanted to get as far as possible away from the coast of Norway and the German planes, so we steered due west for forty hours, then in a southwesterly direction for the next forty-eight hours, before shaping a course for the Faeroe Islands."

They arrived at the islands on the morning of the fifth day, but they were not allowed to enter the harbour of Strumfjord until that afternoon because a German submarine had been seen in the fjord and depth charges were being dropped. That afternoon they entered the port, but could not come ashore for nearly a week while local authorities confirmed that they were not spies. Finally they were cleared to come ashore at Torshavn, the capital of the Faroes. They had the vessel hauled and a fresh coat of copper paint put on the bottom while the Norwegian consul helped Ottar get entry papers for Canada.

On her first attempt to leave for Canada the *Kaare II* ran into a storm and was forced back to shelter in the harbour at Vestmanna. A week later, on July 3, 1940, the family once again started on the long journey across the North Atlantic. Their Norse ancestors had made a similar passage in their long boats, but they only had the sea to worry about. The Noviks were worried more about enemy planes than the weather. On the fourth day a strong northwest wind prompted them to raise the sails to gain a little more speed. This lasted three days and by the time the wind fell they were among the icebergs off Newfoundland.

Haakon Novik recalled years later: "Late in the evening of the ninth day out from the Faeroes we spotted a steamer up ahead. We blew our whistle and blinked our lights. The steamer stopped and when we came up alongside we could read the name *Braker* of Oslo. An officer came out and Ottar hollered to him, asking if he would be kind enough to give us our position. 'Hey, what the hell are you doing here?' bellowed the officer. So Ottar told him and the mate gave us the information that we wanted. We were on course and just 110 miles from St. John's."

Of course, the *Kaare II* didn't carry the charts for St. John's, so Haakon sat out on the bow with a weighted wire that they used to "feel" for herring in the days before echo sounders. Using the wire to take soundings, they felt their way into the harbour.

Newfoundland was not part of Canada in those days, so after five days the Noviks put to sea again, this time bound for Sydney, Nova Scotia. In Sydney they were welcomed by the local residents as a curiosity, but the crews of a number of Norwegian ships in port really took the brave little boat and her passengers to their hearts. Thousands of miles from a homeland occupied by enemy soldiers, the sailors welcomed the familiar sight of one of their own fishing boats and they provided fuel and gear to the *Kaare II* to help the Noviks establish themselves in the fishing community of Canso.

The family's funds were short and the men were fishers, so once a small house had been located ashore, the refugees turned their hands to what they knew best. They had left their seine dories and gear behind; otherwise the seining of herring might have been introduced to the East Coast thirty years earlier than it actually was. Instead the Noviks began longlining with some Norwegian-made gear they had picked up in Newfoundland.

"Longlines? The local people didn't understand what we meant," recalled Haakon Novik. "The only method of fishing they knew was dory fishing, and they just could not understand how we were to haul our lines by means of a 'gurdy'. They assumed that we intended to haul 'em by means of the gypsy-head on a deck winch. They didn't understand the use of the sheave. We tried to explain, but they just shook their heads and shrugged their shoulders, quite sure, no doubt, that we were a crazy lot.

"Fishing was good, really, but the prices were so low that no money could be made; 1 1/2 cents per pound for codfish, 3 cents for haddock. We ran out of fuel oil, so took our twenty-odd empty barrels on board and went in to Halifax and fuelled up from Norwegian tankers. And we got more than just fuel and lube oils, we got ropes and lines, some paint and food too. If it had not been for this help from Norwegian ships, we couldn't have operated."

But even with this help, it became clear that there wasn't much future for fishermen on Canada's East Coast. On a chilly March 2, 1941, the *Kaare II* left Halifax bound for the Panama Canal and Vancouver. As they headed into American waters a light southeast breeze increased in velocity until they were bucking heavy seas. Ottar took the boat inside Cape Cod in order to run down the Cape Cod canal. Once into the sheltered water they were approached by an American patrol boat. Their fame had preceded them: the crew wished them good luck and left. Soon they passed into open seas again, and the barometer continued to fall. Eventually they were forced to seek shelter in New York harbour.

Foul weather followed them all the way down the coast. One evening the skipper had just turned in when he was awakened by the rising seas, which brought him back to the pilothouse to fight a raging storm. The little boat pitched hard but never faltered. When they arrived in Cuba they learned that the storm was in fact a full-blown hurricane.

The storm blew itself out and the rest of the trip to Panama was uneventful. They arrived in the Panamanian harbour of Colon on March 26, 1941. The pilot assigned to take them through the locks was a Dane, married to a Norwegian, so the passage went smoothly, although the fee of 31 dollars for their 31-ton ship further depleted their limited cash.

The trip up the Pacific Coast took another month and the *Kaare II* arrived in Vancouver early in the morning on April 18, 1941, to begin a new phase in the life of the Noviks. In the summer of 1941 they used the vessel to pack salmon. After the salmon season they longlined dogfish, then joined the Norwegian-Canadian halibut longline fleet. In this role the *Kaare II* went on to earn herself and her family a place of note in the BC fishing community.

In 1951 Ottar Novik built a new boat in the style of the West Coast double-decker seine boat longliner. But he kept the name *Kaare* and had the design for the heavy-timbered stern typical of the West Coast boat modified slightly to reflect the fine-sterned Norwegian tradition. Haakon went on to become a skipper; his last boat before his retirement was the *Brooks Bay*. When Ottar retired he sold the second *Kaare* and she was re-

The *Kaare II* as she looked in the 1960s shortly before she was lost with all hands in Hecate Strait.

named the *Istra*. The original *Kaare II* continued to fish halibut for other owners, until in 1964 the little boat that had escaped the Nazis, crossed the Atlantic, visited New York, survived a hurricane and brought a family to a new land, was lost with all hands in the treacherous waters of Hecate Strait.

"That," wrote skipper Novik at the time, "was the end of the good old motor-knutter, *Kaare II*, the seaworthy fishing boat that had taken us safely, over big and dangerous waters, all the way to another continent on the other side of the globe."

BOATS GOT BIGGER

Seine boats that were built in the 1950s, especially the larger ones, were very different boats than the ones Wallace Fisheries brought up to Barkley Sound just before World War One. Those early seiners had the broad stern and forward engine of the later boats, but they had smaller, lighter winches and deck gear, small wheelhouses and no flying bridge, and both the galley and the crew quarters were below deck. At least one of those boats, the *W. No. 4*, was still working in the 1950s, but she had a new, much larger cabin with the galley moved out of the fo'c'sle and located aft of the wheelhouse, which also contained a small accommodation for the skipper and his wife. In the 1960s the *W. No. 4* had a drum installed in place of her seine table and this further raised her centre of gravity. In 1980, after sixty-eight years of successfully navigating coastal waters, she rolled in a tide rip in Seymour Narrows and was lost.

The changes to the *W. No. 4* reflected changes to the design of new boats over the years. The addition of controls on the top of the wheelhouse gave the skipper much better visibility for spotting fish. In the late 1930s, some large boats were built in Vancouver, several of them at the Boeing Shipyard in Coal Harbour. These included the *BCP 50* (ex-*Van Isle*) in 1936, *Adriatic Star* and *Snow Prince* in 1937, *BCP 52* (ex-*West Isle*) in

The *Adriatic Star* was built in 1937 at the Boeing Shipyard in Vancouver for partners Carr and Katuic. She was one of the first of the bigger boats built to pack pilchards and herring from grounds not so accessible to packers and scows. A few years later boats routinely were built with top wheelhouses, in what came to be called a "double-decker style." A top wheelhouse was added to the *Adriatic Star*. (VPL 24117)

1938, and the *Bligh Island*. The first four of these boats had registered lengths of 66 feet and a length over all of 75 feet, with 20-foot beams. The *Bligh Island*, and her sister ship the *Midnight Sun*, built in 1938 at Pacific Salvage Company, were larger, with a registered length of 71 feet and an LOA of 80. These boats were designed for their Canadian owners by the American naval architect T. Halliday and reflected increased size in the American pilchard fleet. The big boats packed about 125 tons of pilchard: packing capacity was becoming increasingly important, as the pilchard were no longer coming into the inlets on the West Coast. To get them, seiners had to go farther out to sea where it was more difficult to use scows for loading and carrying the fish.

The Boeing Shipyard in Coal Harbour on the Vancouver waterfront was owned by Bill Boeing of the Seattle aircraft firm and was noted for big, heavy seine boats. (VMM)

The economic needs of the fishery dictated the increased size of seine boats, along with the extension of the fish hold forward under the cabin. The engine moved forward and the crew accommodation moved up into the deckhouse. The wheelhouse moved up from the main deckhouse to the cabin top and included a small captain's cabin. Quarters were luxurious compared to the early seine boats, where the crew shared the forward half of the hull with a fuming gasoline engine, the cookstove and their mates' dirty clothes. The design of these new boats became the norm for all boats intended for herring or pilchard fishing, though smaller boats with the crew quarters in the fo'c'sle continued to be built for salmon fishing.

The ongoing competition between American and Canadian seiners to intercept Fraser River salmon before they reached the other's nets also contributed to the growth in the size of seiners. The seine boat has been used as a weapon in the international politics of salmon management by both countries. Some of the first US salmon seine boats were built to intercept Fraser-bound sockeye in the San Juan Islands. By the 1920s significant catches were being made by American fishers on the high seas west of the entrance to Juan de Fuca Strait where they were not subject to restrictions. Following negotiations between American and Canadian officials, a 50–50 division of the sockeye catch was agreed to in the 1937 Sockeye Treaty. Pink salmon were not covered by treaty, however, and between 1945 and 1951 the Americans took 70 percent of the pinks bound for the Fraser. Then Minister of Fisheries Jimmy Sinclair decided that the best way to bring the Americans to the bargaining table was by getting Canadian seiners to go off the coast and catch the fish in Canadian waters before they entered American waters. Canada had fishers who were anxious to get at the fish, but in spite of the wartime subsidy, there continued to be a shortage of seine boats large enough for this type of fishery.

"I told Jimmy Sinclair that we should have bigger boats on the coast if were going to fish outside," Charlie Clarke recalled. "With the dropoff of the pilchard fishery after the war, the Americans in Seattle had

BOATS GOT BIGGER

some big boats for sale. Sinclair looked into it and arrangements were made so that boats could be imported with a 20 percent duty to be refunded after one year. Ritchie Nelson told me to go down and pick out some good boats. I went down and got three. One for myself and two for the company.

A big double-decker fishing in the groundswells off San Juan in the 1950s. It was for this fishery that the Canadian government allowed the import of American pilchard boats. (PAC PA-146256)

"Those Americans are clever fishermen, but they put 110 DC electrical systems in their boats instead of 32 volts. This 110 is hot stuff and three or four of the boats burned up with that. I named my boat the *Veta C* and she burned up on me, right down to the engine room and the stern was still partly there. We had been fishing herring at Nanoose Bay. Nobody was hurt, there was a packer close by, but we lost all of our equipment. She had automatic fire extinguishers but they didn't help.

"I took her up to Mercers Shipyard in New Westminster. I had her insured for $100,000. Mercers bought her from the insurance company, then talked to me about rebuilding her. So we rebuilt it my way. I had them raise it up. We built from the waterline up of yellow cedar. It cost $100,000 to rebuild with the same engine. I didn't get the 20 percent refund on the duty because she burned in the first year.

"She's a good boat now. I renamed her the *Western Producer* to change her luck. She doesn't look anything like she did before the fire. She

Louie Percich's *Marinet* slides down the ways at Menchions Shipyard in 1947. (FT)

had been built with a raised poop deck aft, hollow in the centre, so I pushed the deck right straight through and raised the fo'c'sle up. When I bought her the Americans told me she would pack 150 tons of fish. I loaded her with herring and came into the river. When she hit the fresh water I had four inches of water all over the deck. Ritchie saw her and thought she would sink. When we pumped the fish out, it was only 120 tons. But now, since I raised her up, we took, from Burnaby Narrows in the Queen Charlottes to Port Edward, twelve full loads that averaged 150 tons, with the seine on and the power skiff on top of the seine. The only vessel that ever left that way. Even your new steel vessels like the *Milbanke Sound*, they always took the seine off. But the *Producer* could do it and still no water on deck.

"That was with the old engine. Now that they have put a lighter Cat engine in she'll only pack 130 tons. That's the trouble with the seine boats. They were designed by the architect for heavy-duty engines. Now those are changed for new, lighter engines and a big heavy drum is put on the stern. They end up top heavy and stern heavy. That's why we lost so many on herring. But if you live on a boat for a while you know the feel of it. You know that it's good. You just know it, that's all. If it's topsy-turvy, put some ballast in the bottom of her."

Sinclair's strategy worked. By 1955, the Canadian share of the convention area pink catch had grown from 30 percent to 47 percent. In 1957 the Americans and Canadians signed the Pink Salmon Protocol providing for the same 50–50 sharing of the catch that was agreed upon for sockeye.

The increased Canadian and American high-seas net fishery raised concerns, however. By 1956, 35 percent of the Canadian catch was taken in Juan de Fuca Strait, up from only 3 percent in 1944. The Surf Line agreement of 1957 limited the net fishing from expansion out into the eastern Pacific Ocean in order to allow for effective management. The Canadian boats taking part in this fishery were virtually all the big "West Coast" seiners, a number of which were the imported US boats.

Not all of the big double-deckers were from the US. Several yards, notably Harbour, Bensons and Menchions in Coal Harbour and Mercers Star Shipyard in New Westminster, were building throughout the late 1940s and the 1950s. In the late 1950s, Sam Matsumoto also built several fine double-deckers in his yard on Burrard Inlet at Dollarton.

Dick Anzulovich and Louie Percich both had double-deckers designed by Robert Allan and built in Coal Harbour yards. In 1947 Louie Percich had a 60-foot single decker, the *Marinet*, built at Menchions. He fished it so well that Ritchie Nelson offered him a $100,000 loan, secured with a handshake, to build a double-decker at the same yard. The *Marinet* was a little small for herring. Louie's new boat, the *Nanceda*, launched in 1950, was 78 feet. He fished salmon and herring with this boat. Then, in the 1960s, with the herring over-fished and closed down, Louie took the *Nanceda* south to fish tuna. This led to an offer to skipper a large steel tuna boat out of Montreal and Louie sold the *Nanceda*. She was owned by BC Packers and skippered by Norman Gunderson in 1987, when she apparently sprung a plank as she came out of Barkley Sound with a load of herring. Within thirty minutes of the alarm the crew had been taken off by another boat and the *Nanceda* was gone.

Louie Percich had the larger *Nanceda* built in 1950 to allow him to pack more herring. (FT/CI)

At the time, Louie commented on the boat he had named for his daughter Nancy and his wife Edna. "A boat is just a piece of equipment, a bunch of nails and a few planks. But what matters is if she is a good sea boat. One time on tuna I was about one hundred miles off the coast of California. The weather report said, 'Punta Arenas: wind northeast sixty, sea phenomenal.' We were out there for one week. Every day a Coast Guard plane came and circled us to see if we were still there. Finally we had to go in, but only because we were out of water. She was a good sea boat!"

Louie's partner, Dick Anzulovich, also bought a small seiner after the war. His was the *Rainbow Queen* (ex-*Howe Sound*) which had been seized from a Japanese Canadian owner at the beginning of the war. In 1952 Dick had the *New Queen* built at Harbour Boatyard to a Robert Allan design. He named her in recognition of his previous boat and the newly crowned Queen Elizabeth. "Ritchie Nelson said to me, 'Dick, you need a bigger boat.' I had no idea to go to a bigger boat. Ritchie and Charlie went to Pete Storness at Harbour Boatyard. He got paralyzed and never did see the launch, but he was a great boatbuilder," Dick recalled years later.

For some time it was believed that double-deckers were too big for drum seines, but heavier hydraulics and the addition of a bow thruster have made the drum standard on all Canadian seiners. They are used in Washington State but not in Alaska where they are outlawed as too efficient. (AHB)

The main reason companies encouraged fishers to build larger boats was to assure a steady supply of fish for the processing plants. But it was also part of the strategy to keep fishers in debt to the company so that they were not free to take their fish to the competition. Many fishers were held in debt to the company simply by the annual maintenance cost on their boat and gear. More aggressive fishers had to be encouraged to build new, bigger boats in order to keep them obligated.

And don't let's forget the ego of the individual fisher. The size of a boat was very much a statement of the owner's ability as a fisher and his standing in the fleet. It was all part of the postwar mentality in British Columbia. Whether it was dams or highways or buildings or seine boats, bigger was better.

A GOOD MAN LOSES A GOOD BOAT

The Thompson family of Gabriola Island has been fishing in BC waters for about seventy years. Randy Thompson started off in the 1920s in a converted sailboat bought by his dad, a Nanaimo coal miner. His wife Dolly goes back even further. Her grandfather, Portuguese Joe Silvey, fished commercially out of Reid Island and in 1872 was granted exclusive fishing rights to the waters from Sansum Narrows at the western end of Satellite Channel to Dodd Narrows at the western tip of Mudge Island. Her father, Domingo Silvey, talked Randy into converting his 36-foot double-ender, the *Midget*, into a table seiner in 1931, launching his successful career as a seine fisher. So is it any wonder that the Thompsons' five sons all got involved in the fishing industry?

Of all the boats Randy owned over the years, the biggest was the *Key West II*, an 85-foot double-decker built in the US for the pilchard fishery. As well as fishing her, he logged 100,000 ocean miles doing research

The US-built *Key West II* was about as big as seine boats get on the West Coast. She symbolized Randy Thompson's well-earned reputation as an innovative and successful fisher. (Thompson family)

charters for the International Pacific Salmon Commission. Still, each year when the seine openings began, Randy and his eight-man crew returned to fish salmon and herring. In 1958 they made a record herring set with the boat, 1600 tons at Head Bay in Nootka Sound. There was so much fish that Randy had to give away 600 tons that his packers couldn't carry.

And then, in 1961, he lost her.

"It was kind of a freak accident," he explained later. "Every year I used to take a bunch of Christmas trees over from here [Gabriola Island] to Steveston for the BC Packers office. Instead of going over by car we took the trees over on the boat. Then I went to Seattle for a fishermen's meeting, and stayed overnight. When we came back they dropped me at the Imperial plant in Steveston. I went down to the boat and the watchman cast my lines off for me.

"My brother had cleaned the [injector] nozzles on the big heavy-duty 16-ton Enterprise that she had in her. When he set them he set one a little too low or too high and it was missing on one cylinder. With the big cylinders, it vibrated quite a bit. So I jogged down the river with a fair tide. When I came out I headed for the beacon here and put it on the mike [automatic pilot]. I was just running slow, about two or three miles an hour. I looked all around and there was nobody in sight.

"I went down to the engine room and I reset it on the fly and then I got it too high or too low, one way or the other. So I reset it again and that put me down there for a half hour. There was nobody steering. I didn't notice this tug coming from Howe Sound and he caught up with me. Well, he was the overtaking boat and I had the right-of-way really, but it was noon and the skipper was down below having lunch. There was a deckhand on the wheel, a young kid. They were doing about 9 knots and the kid thought they would get by me, and he did go by but they had two thousand feet of cable out.

"I just came up on deck and there was this barge coming at me. It scared the hell right out of me. I ran up top and cranked her wide open but they hit me. I goosed it and pushed the barge around and got clear of it. Then I ran down on deck and pulled the hatch covers off and looked down and there was no water. I looked in the engine room and there was no water so I thought 'God, I'm all right.'

"I got on the phone and contacted the skipper of the tug and told him I was going back to Celtic Shipyard [on the north arm of the Fraser]. I started back and there was a little swell running. I opened the old thing up. Now, she had big thick guards on the hull. When the barge had hit, it had lifted the guards up and split the hull open just above the water line and I hadn't seen it. So when I headed back I drove the water right in through the hole under the guard. It didn't take long for a lot of water to come in. When I realized that I was starting to go down by the nose I ran back down to the engine room. The engine was deep in water. I tried to dive down and open her big pumps up but I couldn't make it. I phoned the tug again and said, 'I'm coming back. I'm sinking.'

"I made it alongside the tug and they put huge ropes across. They said, 'We've got two big pumps.' They put the big pump on the fall and they dropped it on my deck and then they put the little one on. The big pump had a 4-cylinder gas engine. It got the ignition wet and I couldn't

Dolly Thompson has shared the feelings of many fisher families who have had to wait home and wonder why the boat was late getting in. (AHB)

get it started. I finally got the little one started and it wouldn't pump. They wouldn't come on my boat and give me any help except one of the deckhands who came over and gave me a hand.

"There was a Standard Oil tanker, not a real big one, I guess about 200 feet, and he heard us on the phones. He said, 'I've got some tremendous pumps. I'll come alongside if I can be of any help.'

"I said, 'Good! Fine, come on.'

"So he came on and he was excited and in a hurry and that oil tanker just hit my boat and squashed her up against that tug like a sandwich and down she started to go. The only thing you could do was grab the axes and chop the lines and let her go.

"One thing after the other. But I just looked at her this way. Maybe it was meant to be. I've been out and done 100,000 miles out there on the Pacific Ocean. I rode, lots of times, winds up to 100 miles an hour and I've had some rough trips. When you get a thousand miles offshore, you know, there's no anchorage. So I just looked at it that way."

Lowfloat Log in the Forefoot

Roger Skidmore, a Comox fisher, had been fishing halibut for eight years in the spring of 1987. For the last two of those years he had worked the little 41-foot *North Foreland*, which was built in Prince Rupert in 1934, without mishap. But that April 1987 opening was very different.

"I left for the grounds early so that I would have lots of time to pick up bait in Sointula and get to my fishing area north of Prince Rupert. I had picked up Derek Cox—he and his wife run the PetroCanada marine fuel station in Thulin Channel—to crew with me.

"There was just the two of us on board as we toddled along up Lewis Channel at about 1:30 on Monday morning of April 27, when whacko! We hit this big log and things turned into a mad panic. We were just going by Tea-kerne Arm where the big booming ground is, and I guess a big lowfloat log had come out of there. We smacked her right dead on in the forefoot. It took the forefoot out and the water came in. My pumps wouldn't keep up to it, so I tried to beach it, but in Lewis Channel there aren't any bea-ches. It is pretty straight up and down along there. I managed to tie the bow up to a tree and set the stern on a rock. It was about an hour be-fore high water when I got to the beach. I sent out a May Day; Campbell River Coast Guard tried to get someone to help me but the few that offered to help didn't have any pumps. The big Coast Guard boat was in Victoria for a refit, so they sent the little 38-footer out.

"We still had the bow sitting on the rocks when they came, but it was about an hour

Roger Skidmore points to the bow stem damage on his boat that nearly cost him a halibut opening. But by the time this photo was taken he had delivered a good load at a good price to a buyer in Sointula. He is headed home to Comox and the world is wonderful! (AHB)

after high water. The main deck was awash but the water wasn't coming in the galley door yet. There was air space in the lazarette keeping the stern afloat while the bow was on the rock. It kept slipping down the rock all the time. The water had covered the whole engine, the batteries, everything. The 12-volt system was still working, but when you stopped talking to the Coast Guard you could hear the bubbles coming out of the battery bank. The 32-volt system packed it in shortly after it went under the water.

"I was a bit upset that the Coast Guard dispatch took an hour to respond, but the crew on the Coast Guard boat were really great. They put two 3-inch pumps on board and towed me to Lund—it took about five hours but they had the pumps on board the whole time. We put the boat up on the ways at Lund Marine, where Mark Sorenson did all the mechanics on it to bring it back up to snuff. The water hadn't got into the engine-head so it didn't ruin all the valves and stuff; it just needed flushing out. We cooked the starter and they put it back on so that it worked fine. Billy McKee from over in Sevilla Island Boat Works was the wood man. He came over and put a new forefoot in her and caulked her all up. Actually he put two forefoots in, one to make up the damage caused by the log taking out part of the bow stem too. It only needed a couple of little planks—one on each side just below the water line—but it's inch and three-quarter planking and it would be hard to bend. So he took it back to about where the guard starts.

Long admired as one of the best-looking boats on the coast, Walter Carr's seiner *Waldero* was being operated by another skipper when she was hit by a freighter in Johnstone Strait. The foam insulation and solid construction kept her from sinking, but now it is time for estimates on just how much it will cost to repair the damage. (AHB)

"We went up on the ways on Tuesday and we were back in the water on Wednesday afternoon. I had to pick up my other deckhand in Rupert, and we worked day and night. On the way up we had some small problems. We had to replace the 32-volt battery bank. I tried the hydraulics and they had water in them so that had to be flushed out. The fishing opened on Saturday at noon and we got up there on Sunday but it had been blowing a gale and there wasn't anyone fishing anyway. On Monday it was still blowing pretty bad but we took off and got the gear out on Monday afternoon, so basically we began fishing on Tuesday. We found a few halibuts and made enough money to pay the bills off. For the whole time we were fishing the pump was coming on about every three minutes for fifteen or twenty seconds. It wasn't leaking so much that it was a panic, but you had to keep your ear tuned even when you were sleeping. It is hooked into the sink and you can hear it gurgle when it comes on so I could hear it. I was sort of like a dog sleeping with one ear open."

Mere Words Do Not Express It

In late February of 1961, I took a small part in the search for the seiner *Northview*. She had gone down in Finlayson Channel when a following sea pushed by 80-mile-per-hour winds lifted her stern quarter until she pivoted over on her narrow bow. At least so the theory goes. No one was there to see her go over and none of the eight crew members survived. Along with the seventy-boat herring fleet, the boat that I was crewing on had just weathered the same storm and had some understanding of the forces involved.

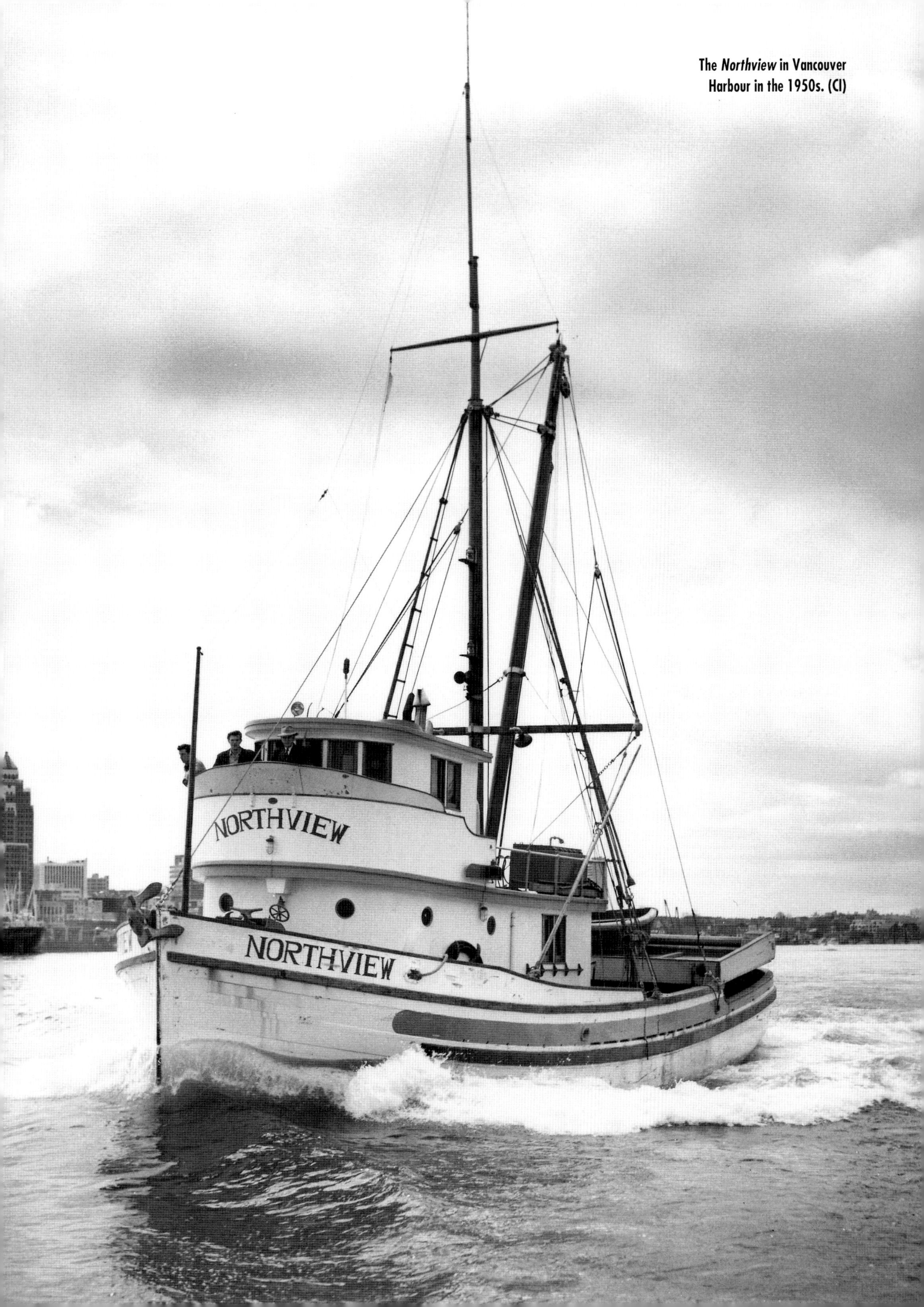

The *Northview* in Vancouver Harbour in the 1950s. (CI)

As we searched the waters that morning, listening to the skippers talking with each other on the radio phone, I felt the last of my teenage innocence of death draining into the dark green depths where I expected any moment to see a water-whitened face staring back. With or without the actual loss of life, the sinking of a boat reminds us all of our mortality. In a report of the incident, made at the time but published in a fishing journal many years later, Captain Charles Fletcher of the seiner *Misty Moon* concluded, "The response of captains and crews of the various boats is beyond praise, mere words do not express it. I am sure that to each man as he left came a feeling of frustration and sadness, saying to himself 'Did I do enough?'— 'Could I have done more?' even knowing that he could not. I know I felt that way and it is with deep sorrow and regret for those left at home that we have to sign our names to this report."

Joe Katnich Sr. came from the town of Crikenica in Croatia in 1926. He became a very successful fisher and owned the seiners *Westview* and *Eastview* before building the *Northview*. He was in the process of turning the family business over to his son Joe Jr. when the *Northview* went down with all hands in 1961, taking Joe Jr. and his brother Mario with it. Left with one son and a daughter, Anna, Joe Sr. sold his other boat. Anna Katnich says that the loss left a hole in the family that never healed. Her father never went back around the boats. He died in 1988. (FT/CI 1036-1&6)

FRED KOHSE AND THE *SLEEP-ROBBER*

red Kohse has been around boats all his life. His father, who worked on the construction of the Empress Hotel, owned and operated the Empress Drive Yourself Motor Boat Company in Victoria's Inner Harbour. As a nine-year-old, Fred was taking fishing parties out in the 22-foot clinker-built boats powered by Hupmobile and Cadillac car engines. This was the era of rumrunners, and young boys growing up on the Victoria waterfront learned a lot about the construction details and operation of specialized boats. Especially if Johnny Schnarr, one of the leading members of the rumrunning fraternity, lived in your house.

In the early 1920s Fred's father moved the whole family north to homestead at Port Hkusam near the mouth of the Salmon River. "They never should have allowed people to farm north of Campbell River," Fred says. "With the money Dad paid for the place he could have bought a place around Saanich. We used to plant potatoes at $3.50 a sack and when we dug them they were a dollar a sack."

The area may not have been productive farm country, but the First Nations people had built a rich and complex culture on the marine resources. It was people from the village at the mouth of the Salmon River who in-

Port Hkusam is still a pretty spot, but when Fred Kohse was growing up he was unimpressed by the notion of farming there. (AHB)

troduced Fred to trolling from a small boat, and he recalls watching as "a seiner named the *Hillside* came in our bay and made a set and got 11,000 humps. It's kind of rocky in that bay but he got away with it. We loaded up this boat that Dad had got from down below. We took the fish up to a cannery owned by old man Stump up at Charles Creek in Kingcome Inlet. I think that was the first time I ever had a look at a cannery. This must have been about 1926.

"I started out gillnetting about 1931. I remember an American troller anchored out in our bay. My dad sent me to row out and invite him to come in for dinner of pigs' feet and sauerkraut. He told us all about the big run of sockeye up at Smiths Inlet. So my brother-in-law and I rowed up to Smiths Inlet from Salmon River. We went to Margaret Bay and asked the manager there, Old Trotter, for a gill-net boat. He should have realized that two fellows who rowed all the way up there have got to have a little gumption. He told us we could have one of the skiffs that was laying there sunk. We bailed this damn thing out and fixed up a little primus stove and rigged the tent over the bow. There were shacks ashore but he didn't mention them to us. I guess he just wanted to see what we were made of. When he came down in the morning it was raining like hell but we were still quite cheerful so he said, 'Well, I'm going to give you guys a net.'

Fred and his brother couldn't afford steamboat passage on the SS *Camosun* to Margaret Bay in Smith's Inlet, shown here, so they rowed the nearly 100 miles. Once there, the cannery manager offered them a sunken skiff like the ones in the centre of this photo. But they soon proved their abilities and Fred went on to one of the most successful fishing careers in BC history. (VCA)

"He made the mistake of giving us a three-year-old linen net and an old flat-bottom skiff. They had the round bottom Columbia River type boats with the sail on them, but you had to be a good fisherman to get one. That old net was rotten. It could catch a few fish up at the head of the inlet where they have slowed down a bit and don't hit the net so hard. But we followed the top fisherman out to the mouth of the inlet. Whenever that fellow hauled his net we hauled, whenever he set, we set right by the end of him. We would see fish hitting our net but we would just get one or two because they were just going right through that rotten old net. Well, that's where Trotter made his mistake. He lost money by giving us that old net. That's why Ritchie Nelson was so successful with Nelson Brothers. He would size up people. He would see somebody who's sort of anxious and wants to go and he would back him a little bit. Trotter wasn't like that. He would come around with a long cigarette and look down at the fishermen like they were the scum of the earth. But we made $110 that year, even though we had to row back down to Salmon River and help put up the hay. Then we rowed back to Smiths for the rest of the season."

About 1933 Fred got his first boat when Logan Schibler of Owen Bay gave him the burned-out hull of a gillnet boat called the *Wonder 2*. Fred rebuilt the cabin and repowered this first boat. A year later he copied the power-driven net drum that he had seen on a boat from Sointula where the innovation of powered net drums was first developed in the 1930s. The boat was less than ideal but it was a big improvement on the rowed, flat-bottom skiff. Fred fished Smiths Inlet for some time before rigging another boat for trolling out on the West Coast and south to Neah Bay and the Deep Hole. Then, while beachcombing in Johnstone Strait, he hit a log and sank the boat with only enough insurance to cover the cost of the new Fairbanks engine.

It was time to get a proper boat. Fred and his young friend took a job logging up in Knight Inlet for Einer Johnson. "I was there 205 days and I worked 203 of them. I got 5 dollars a day clear so I got a cheque for $1000 and I sent it up to the boatbuilder Cholberg. On the farm we used to work from daylight to dark. When we went to the logging camp you started at 8 a.m. and quit at 4:30. We didn't know what to do with the rest of the time. So we used to go and help the boss in the blacksmith shop. One of us from the farm would be turning the forge and the other sledge-hammering and we'd be making chokers and giving him a hand. So he gave me all those yellow cedars that they couldn't sell anyway. He let me have his camp boat, which was the seine boat *Canadac*, to tow the logs down to the sawmill and then use it to pile the lumber on and take it up the Mr. Cholberg. This was a big help. I think anyone who has gone along in his life and had some success has to look back in life and thank people that have helped you along the way. You'll often see that these old-timers will help if they can."

Christian Cholberg, the builder that Fred had chosen at Cascade Harbour on Nigei Island, was an old family friend from Victoria. He had owned a large shipyard in Victoria during World War One, where big sailing barquentines such as the *S.F. Tolmie* were built. It was said that these boats used the lumber from twenty-five boom sections of logs. After the war, with the decline in large wooden ships, Cholberg had moved to Vancouver, then north to Cascade Harbour. With Fred's hard-earned money and lumber, they built the 39-foot *B.C. Troller* and launched her in 1939. Now Fred had a boat that could take him anywhere.

In 1941, he took her north to Prince Rupert with A.C. "Tony" Vick, a clothing salesman from McKay, Smith and Blair. They made sales stops at logging float camps and fishing ports along the way. In February 1942, the boat was rigged for longlining and Fred took Pete Gunderson as skipper for halibut and black cod fishing. In 1943 Lief Tusvick joined the boat and became the second skipper. Fred says that with experienced men like Lief, who knew all the halibut spots, and with crewmen like Nels Johnsen and Martin Gronmyer, he simply couldn't fail. This was Fred's "in-breaker" period, as he served his apprenticeship in the fine art of longlining, including being cook and engineer. Prince Rupert was the halibut capital of the coast and the Norwegian-Canadians were the kings. It was not easy to gain acceptance in this select company, but the life of an

Fred's first new boat, the *B.C. Troller*, cost him 205 days in a logging camp.

"in-breaker" fishing black cod off the Alaska coast soon separates the fishers from the farmers. "I thought myself it was damn tough. The first time we were out there in March and it was snowing and blowing, I started looking up into the bush for some trolling poles. We would fish all day and come in at night around ten or eleven o'clock and put these nets out to catch herring for bait, then get up at two in the morning to haul these nets and shake them out on deck. One morning this Norwegian named Martin started cursing the skipper and it struck me as so funny that this old-timer would think it was tough too." It didn't take long for the owner of the *B.C. Troller* to gain full recognition as a fisher.

Fred's reputation for hard work and fishing savvy grew quickly, so that in 1945 he was asked to take on a very special project. "Vivian Engines and Star Shipyards built the *Arauco* and they asked me if I would take the boat down to South America. I got my crew from the *B.C. Troller*, which I sold in the fall of 1945. We went down to show those fellows how to operate the longline gear and the drag gear we had on there. They

If a man can take a rowboat upcoast to Smith's Inlet, then why not take a seine boat down the coast to Chile? Fred did it in 1945. (PABC 89395)

wanted us to take her down to Chile to show what Canadian fishing boats and gear could do.

"It was paid for by Mercers Star Shipyards and Vivian who put up the engine. The boat alone cost $35,000 and altogether with all the nets and gear it cost about $100,000. We went down to southern Chile at a place called Puerto Montt. It was very much like Prince Rupert. The tides were twenty-four feet."

After a year they sold the boat and returned to Canada. Fred had been impressed with the *Arauco* and now went to Mercers Star Shipyards to have a similar boat built for himself. The *B.C. Producer* was launched in 1947 with a registered length of 57.5 feet and a gross tonnage of 48. She was built for longlining halibut and dragging (trawling) other bottom fish, but she had a basic seine boat design. In 1951 Fred decided to rig her with a table and live roller for seining salmon. During an eight-day halibut opening in August, Fred and his crew made two trips, for a crew share of $2300. The crew of halibut men weren't too enthusiastic about seining for salmon, and with money in their pockets they hung around the pub while Fred worked to ready the boat for a seine opening. After getting the basics of seining worked out in northern waters, they headed south to Fred's old home waters in Johnstone Strait.

The Norwegian-Canadian fishers found hand-pulling a seine net very different from the power gurdy-pulled longline fishing. Traditionally, fishers cut the cuffs off their flannel work shirts to prevent the chafing caused by salt buildup. But when they got to pulling the seine net in Johnstone Straits with a broken live roller, one of Fred's crew of halibut men summed up the general frustration: "You don't have to cut your sleeves in Yonson Straits," he lamented, " 'cause your arms stretch out enough!"

Fred Kohse had mastered gillnetting, trolling, longlining and dragging, but he says that he was never really good at salmon seining. He continued to take the *B.C. Producer* after halibut during the spring openings. With a capacity for 66,000 pounds of iced halibut, she could fish farther afield than his previous boat. He began to explore out toward the Westward Islands; a favourite ground was around Chirikof Island. In the winter months, Fred and his crew rigged the boat for dragging, with otter boards and trawl nets. They then headed out to the shallow waters of Hecate Strait off Bonilla Island where Fred saw the herring seiners make their catches of hundreds of tons, load themselves and the other boats in their pool, and head back to the reduction plants at Port Edward, Butedale and Namu.

Fred decided that this would be a good fishery for him. In the early 1950s, before the power skiff had been introduced, smaller seine boats were employed to "tow off" the larger seine boats. This was done by towing on a line attached to a bridle on the seiner's side which kept the boats from getting tangled in their own nets. Fred tried to get work for the *B.C. Producer* as a tow-off boat, but he was rebuffed. This led him to think of a seine boat that was large enough for herring seining and with the packing capacity to make halibut trips to the Bering Sea profitable.

For the construction of his new boat, Fred went back to the Mercers Star Shipyard in New Westminster. For the design he turned to naval architect William Garden. This was before Garden had moved from Seattle to the splendid isolation of his island off Sidney. Fred was impressed with the large beam that the Americans were building into their Alaskan seiners. (These boats were limited by regulation to a length of 58 feet and could get larger only by adding width.) Fred particularly admired an American boat he had seen, the *North Pacific*. It was those lines with which Bill Garden started to work.

The resulting boat caused quite a stir among BC fishers because of her large capacity. Dick Anzulovich's boat the *New Queen* had a registered length of 67 feet and a beam (width) of 18 feet. Fred's new boat was a foot shorter and two feet wider. The greater beam and fuller design gave the boat 74 registered tons, compared to the *New Queen*'s 61 tons. For the fisher, especially when packing halibut all the way from the Bering Sea, this meant a considerably larger payload.

Dick Anzulovich's *New Queen* was typical of the postwar big boats in BC, but Fred Kohse wanted maximum packing capacity for his Bering Sea halibut trips so he went to designer Bill Garden for the innovative design reflected in the *Sleep-Robber*. (FT/CI 1390-3)

The transom stern on the new boat was also a departure from the traditional timbered sterns. On Fred's boat the horn-timber, which carries the weight of the stern down to the keel, extended only up to the water line where a nearly flat, planked transom took the place of the more usual raised and timbered stern. This meant that as gear or fish were added there was more immediate buoyancy achieved on the stern, which reduced the "hogging," or tendency of a boat to bend up in the middle and down at the bow and stern. The transom stern also was supposed to be stronger, less costly to build and cheaper to repair.

Fred named his new boat the *Sleep-Robber*, not simply for the long hours that he expected to work her on the grounds, but primarily in admiration of Mao Zedong's guerrilla fighters who had earned the name by stealing guns from sleeping Nationalist soldiers. The name expresses a bond with rebellious farmers who had left their farms for a new life.

Before Fred began building the *Sleep-Robber*, he approached Ritchie Nelson about herring fishing. Ritchie promised to supply a herring seine and power skiff. At the launching he apologized to Fred that the other skippers would not accept him in any of the pools of boats that generally worked together. In those days sets of several hundred tons were common and no boat packed more than 150 tons. The pool was a way to load several boats from one set. It also averaged the risks of one or two boats going off in search of fish. But Fred assured Ritchie that he didn't expect to be carried by anyone while he learned to fish herring.

Just as he had when he fished halibut, Fred hired an experienced older fisher to come along. "I took old Eli Skog. We never caught no fish

FRED KOHSE

the first week. We had those big drums that we used for dragging. We were fishing off Deepwater Bay and we must of made fifteen sets for 110 tons of herring. On the way down to Steveston we stopped in at Ballenas Channel. We had one of the first echo sounders in the skiff. We were quite modern that way. The skiffman saw a school of fish and we set right around him. We hauled our net up and we had only two brailers of fish. I said, 'Eli, I think we are pursing too fast. The end of the net is digging into the mud and when we're pursing she is lifting up and we are forcing the fish out.'

"'Ya, maybe you're right,' he says.

"So the next Sunday, when we were coming out into the Gulf, I was running along and I saw quite a good school of fish. Johnny Dale was ahead of me and he was calling one of the boats in his pool. So I knew he had the school of fish. So I said, 'Let her go right now.'

"When we started pursing I ran down to the engine room and took the screwdriver and slowed the Cat [engine] right down so that she was just barely ticking over. When the rings came up the engine stalled because that is the biggest lift. But then a bunch of foam showed up and we had 225 tons of fish. So after that we did quite well."

Fred says that the fisheries in which he was most successful were those where he had the opportunity "to work under good captains like I

The *Sleep-Robber* is rigged for halibut longline fishing. The poles in the rigging are weighted on one end and flagged on the other to mark the end of the skates. (PABC-E9030)

did when I broke into the halibut under old-timers like Pete Gunderson and Lief Tusvick and herring seining under Eli Skog. On the other types of fishing, like trawling and salmon, I never had the experience that you can only get working on deck under a top skipper and with other well-broke-in crewmen. That's the right way to start out. You might make it starting right out in the wheelhouse, but you'll have more misery."

The herring nets were massive. Fred fished a 28-strip cotton herring seine that would touch bottom in 42 fathoms of water. Some fishers liked to use 32 strips and had to be careful in anything under 50 fathoms.

The *Sleep-Robber* cost $153,000 to build. Fred sold the *B.C. Producer* for $35,000 and his house in Prince Rupert, but still owed $90,000 on the boat. Because of the versatility of the boat, however, he could fish nearly year-round with herring, salmon and halibut. The old-time halibut fishers, with their deep, narrow-hulled schooners, had nothing but disdain for the combination seine boats that were becoming increasingly popular on the grounds. They emphasized the difference when calling on the radiophone by pointing out that they were schooners: "Hello the *Sleep-Robber*, this is the schooner *Nord*." Yet Fred remembers coming down from a Bering Sea halibut trip with a 70-mile westerly gale building massive following seas. They had slowed from the usual 1200 rpm to 800 rpm when they came on the halibut schooner *Pacific* hove to. Fred called the skipper to check if things were all right and was told that, with the following sea and a load of halibut forward, the stern would not pick up and the seas were breaking over the boat. With the broad seine boat stern, the seas would pick up the *Sleep-Robber* and push the boat along.

But constant dangers follow a boat under such conditions. In the spring of 1964 Fred was returning across the Gulf of Alaska from a Bering Sea trip. "I think one big wave came down on us like that. We were running on a slight angle, before the swell. There was almost no wind, I don't think it was blowing any more than 15. We were going, not full speed, but quite nicely. Rick Johanson, who later had the *Bravado* and was lost, was on the wheel. The boat had been on the mike [auto pilot] but she had been sheering off a bit and Rick had taken over. I was lying down. All of a sudden there was this horrendous crash. I thought for a minute it was an aeroplane had landed on us. There was a breaking of lumber and smashing, I couldn't figure what the heck was going on. I looked down and there was nothing but wreckage all around. The swell had come down on the starboard side. It smashed twenty feet out of the side of the house. The heavy plywood was all broken up in little chunks. It lifted the cap off the rail and sprung some of the planking above the water line. We lost our steering when the hydraulic

The *Sleep-Robber* stands by to brail herring in Loredo Sound in February 1961. The dark hardwood on the bow protects the planks from the anchor when it is being raised, while the hardwood amidships protects the side of the boat when halibut longline hooks are being retrieved. (AHB)

lines were torn out with the cabin wall. In the engine room the engine was half under water but the breathers were out so it kept running. We had one of those automatic inflating life rafts on top of the cabin, about four-teen feet up; it was torn off and inflated, but it was still attached to the boat by a line.

"When the wave hit, Charlie Madden, Phillips and, I think, Ed Bowen were lying down in the living quarters. Jacob Jacobson, the cook, and Jimmy Riley were in the galley. The galley windows were covered with plywood storm windows but the wave ripped off the whole wall and broke out through the galley door on the back of the cabin. The force of the water broke the lock on the door. There were glasses in a rack high on the galley wall. Afterward we found these to be full of water. The water hit with such force that the four-inch aluminum table support was bent right over. When the water went out of the galley, Jacob was left with two cracked vertebrae and Riley with his leg hurt.

"No one was hurt in the crew's quarters, but all of their clothes and bedding were washed down into the engine room. The clothing kept plugging the intakes on the pumps so I had to keep diving down to unclog them. Luckily the auxiliary was located high enough so that we had this to support the main engine pump. We lost electrical power for a while when the lines were torn out, but we managed to get it going again.

"We threw the clutch out and lay there drifting while we got to work on the boat. We fixed the steering and spent ten hours pumping out the water. It took a few hours before we knew we were gaining on the water and that we would be OK."

When the *Sleep-Robber* limped into Vancouver some time later, the newspaper featured a photo of Fred Kohse standing in the gaping hole in the side of the cabin, looking justifiably pleased with himself.

Fred sold the *Sleep-Robber* in 1971. He had already built his first steel boat, the 96-foot *Eastward Ho*, in 1969. The time of change from wood to steel, aluminum and fibre-glass construction had come. Old-growth timber was becoming hard to get, and the cost of wages had gone up at the same time as improved welding techniques made steel and aluminum the popular choices for seine boats. In 1972, Fred built the still larger *Southward Ho*. He went on to explore other waters, taking his boats eastward to the Atlantic for herring and south-ward for tuna. He fished tuna in the Philippines with giant seine nets 500 fathoms long by 125 fathoms deep. More recently Fred has travelled to Brazil as a consultant for the Canadian government. In 1993, his *B.C. Troller* and *Sleep-Robber* were still fishing.

Fred Kohse's boats were often seen at Kelsey Bay while he went up to visit his mother in the store that she kept at the head of the wharf until well into her nineties. On the occasion of this 1988 photo he had driven up with his good friend Jack McMillan of J.S. McMillan Fisheries. (AHB)

Fred Kohse could have fitted a half dozen of the rowboats that he took to Smith's Inlet into the power skiff on his last boat, the *Southward Ho.*

THE ERICKSONS OF HARDWICKE ISLAND

Fishing gets in the blood, they say, so it is not surprising that there have been many second-, third-, even fourth-generation fisher families on the coast. One of these "dynasties" is the Erickson family, which lived on Hardwicke Island opposite Kelsey Bay for many years.

The first Erickson to take to the water was Gus, a logger on Thurlow Island in the 1920s. Gus used to row across Johnstone Strait to Kelsey Bay in his little boat, the *Erin*, for supplies. Along the way he began trolling for salmon and on one trip caught ninety-six coho, which he sold at great profit. It dawned on Gus that this was a lot more sensible way of making a living than chasing around a sidehill. It was time for his son Norman to start school, so in 1926 the family took up fifteen acres of land on Hardwicke. They planted fruit trees and a garden and Gus went trolling for salmon.

Once he had learned his multiplication tables, young Norman quit school to join his dad. Like Fred Kohse, who had grown up just across the strait at Port Hkusam, young Norman had watched the steady marine traffic up and down Johnstone Strait and had no intention of sitting ashore. He bought his first boat, a 26-footer called the *Passing Cloud*, in 1936 when he was just sixteen. "I bought her off an old fisherman for $450," Norman recalls. "It had a 5-horse Vivian in it and I travelled to Vancouver from Hardwicke Island with my dad on only twelve gallons of gas."

Later Norman bought a 30-foot boat from John Ines which had been built in New Westminster by John Stokkland. He trolled up into Rivers Inlet and out around Egg Island and Watch Rock, catching mostly coho with a mix of springs. Eventually his wanderings took him to Lund where he met and married Alice Larson, daughter of Gus Larson, a fisher and boatbuilder. Gus had a place in Larson's Bay south of Lund and it was there, in 1945, that he and his sons helped Norman build his first boat, the *Sea Star III*. At 38 feet long and 11 1/2 feet across, she was built to collect fish from gillnetters. "The experts said that you couldn't build a boat that big from cedar," says Norman, "but we did and she's still going."

The cedar was cut by Logan Schibler at Owen Bay, who had a reputation for cutting first-class boat lumber. To assure a quality job, all the nail holes were drilled through the cedar planks into the oak frames. "We all worked on it and drilled all of the nail holes with a little crank drill. Mom even drove nails." The total cost of the *Sea Star III* was $5000. The original engine was a Chrysler Crown flat-head six that cost $800. In the seventeen years Norman owned the boat, it needed only one valve grind.

Wes and his granddad Norm Erickson.
(AHB)

In the summer of 1946 Norman chartered the *Sea Star III* to Canfisco. He made a deal with manager George Russell Clarke for 15 dollars a day plus gas, and bought springs from gillnetters in Bute Inlet for 7 cents a pound. In those days logging had not destroyed the streams and the Chinook runs were still strong. Springs from Bute were known for their white colour. Norman recalls that of the 26,000 pounds of spring salmon that he took out of the Inlet, only 56 pounds were red in colour.

Norman and Alice remained at Larson's Bay for a few years, then moved briefly to Heriot Bay on Quadra Island. Norman fished in the summer and logged or worked on the booms in the winter. In the early 1950s they moved back to the family place on Hardwicke Island where they put their growing family in school, just as Gus Erickson had done a generation earlier. The three children—Steve, Ken and Christine—grew up in the bight of

An American boat loaded with cases of canned pink salmon neglected a course change and ended up on Earl Ledge at Hardwicke Island. (Erickson family)

the bay near the Bendicksons' logging camp and the treacherous reef that extends off the shore toward Kelsey Bay. Norman recalls one time the American boat *Commander* ran solidly aground on the reef. She was on her way down from Alaska with a load of canned pinks which had to be taken off so that she could be refloated. Everyone got lots of canned fish, but according to Norman it was of such poor quality that even the cats turned their noses up at it.

In the early 1960s, when it came time for their youngest child, Christine, to start high school, the Ericksons moved to Campbell River. There Norman built a new boat, the *Northern Spy*, from a Frank Fredette design with Johnny Palmer as head shipwright. He fished her until 1977 when he built his next boat, the *Silent Partner*, to a design by his oldest son Steve. The family returns to Hardwicke now only to visit, but in their homes at Campbell River and Comox, and in the lines of their fine boats, the best of the BC coastal tradition stays strong and well.

The *Northern Dawn*, launched in 1961, was Sam Matsumoto's last big wooden boat. Note the 1958 Chevy station wagon getting a ride to Prince Rupert, and the streetlights mounted on the cabin for use in pit lamping herring, a practice that is no longer allowed. The aluminum skiff is a foreshadowing of things to come. (FT/CI 2742-1)

THE BOATS OF
THE 1950s

O ver the years the evolution of seining led naturally to changes in the design and construction of seine boats. During the 1950s two important boatbuilders in the Lower Mainland—Mario Tarabochia at Ladner and Sam Matsumoto at Dollarton—were manufacturing vessels that defined a dramatic change in seine technology.

Old-fashioned table seining had been pioneered in Canada before World War One by the Martinolich family. Between the wars the Martinoliches continued to be leaders in technology and as fishermen. In the 1930s a Sointula fisher named Laurie Jarvis developed a powered drum system for hauling back gillnets. The powered gillnet drum was immediately adopted by almost all fishers on the coast. In the 1940s a number of fishers attempted to develop a similar system to take the back-breaking work out of hauling a seine net. A gillnet is a simple arrangement of corkline, web and lead line, all set and retrieved in virtually a straight line. A seine net, on the other hand, is much more complex. The lead line is shorter than the cork line, and extra web is hung into the net to form a slight bag to help hold the fish while the net is being towed and then pursed. On a table seiner the net is pursed from both ends. All of these factors made the development of the drum seine particularly difficult. But the advantages were great enough that a number of fishers continued to experiment.

Randy Thompson of Gabriola Island used a wooden drum on his little boat, the *Fauna III*, in the late 1940s. At Nanaimo the Kelly brothers were working to perfect a system that could set a net with the complex design of a seine and be pursed from only one end. The design had advanced far enough that in 1952 the Martinoliches were able to have two boats built as drum seiners. They approached Mario Tarabochia, who had worked with them ever since his brother Joe brought him from the family shipyard at Luccinipocillo in Italy to the Martinolich Shipyard in Oakland, California, and from there to Port Guichon, just down the coast from Ladner. For a time the two brothers built boats together under the name Tara Brothers. (The older brother, Joe, who was already in the US during World War One, had changed his name to avoid being discriminated against as an enemy alien.) The two built many boats, including the *Cape Swain* (ex-*Snow King*, ex-*Arctic Queen*, 1927), *Sandra L* (ex-*ESTEP No. 1*, ex-*Robert Ray*, 1929) and *Z Brothers* (1929), Joe Boskovich's *Neptune No. 1* (ex-*Nimpkish Producer*, ex-*Neptune I*, 1930), and Charlie Clarke's *Western Monarch* (1937). Joe Tara built Luigi Benedet's *Ermelina* (1938), Joe Boskovich's *B.C. Lady*

(1939), and John Radil's *Ivana*. In 1940 Mario Tarabochia built the *Splendour* for the Martinoliches. Joe Tara Sr. died in 1943, leaving a fourteen-year-old son, Joe Tara Jr., who apprenticed to his Uncle Mario at 40 cents per hour. In 1951, he helped his uncle build the 43-foot Seattle-designed seine boat *Mary N*.

It was the design of the *Mary N* that they stretched the following year to build the drum seiners *Marsons* and *Marlady* for the Martinolich family. Richard Martinolich, who had Mario build a third boat, the *Mar-brothers*, in 1955, recalls: "The drums were built right into a well in the deck. The boats had a fuller stern to support the extra weight. We didn't know anything about drum seining to speak of, but the only way to find out was to try it. The first year we had one devil of a time with net problems. We hung our net the old way with the lead line five or six feet shorter, but it didn't work worth a darn. We were getting roll-ups all the time. To drive the drums at first we had the rear end off a truck with a chain drive to the drum. We had a hydraulic motor running through the truck transmission. They made expenses the first year but that was about all. In the middle of the first year they rehung the nets and lengthened the lead line. The next year we had problems because we were still using cotton web and manila lines. They stayed damp on the drum and rotted. The Spanish corks were popping off the cork line like hotcakes. It was better when we got nylon because it didn't rot. The first year they had the purse line right through, but they only pursed from one end. The second year they cut fifty fathoms off the towing end and then they cut it a little shorter. We ended up hanging the lead line two to two and a half feet shorter. Now some of them are hanging anywhere from 25 percent to 40 percent of extra web."

The early drums did not win universal acceptance. In the early 1950s, a Seattle inventor patented the Puretic power block for use with seine tables. This device hung on the end of the boom and was driven by rope from the winch on some early models and then by hydraulic motors. The whole net passed through the big shive of the block and the net was piled on the seine table by the crew. Through the 1950s this was the favoured method, but with the advent of nylon nets and the refinement of fishing techniques the drum won out. A crew was able to make two or three times as many sets in the same amount of time with the drum as

The chronology of Randy Thompson's boats included experiments with drum seines on the *Fauna III* in the late 1940s.

with a table seine with a power block. At first it was believed that a drum could only be used on smaller boats but eventually it became standard on even the largest boats. Recognizing the danger of having their already large fleets become more efficient, the United States banned the drum seiners from Alaskan waters, although they are used in Washington State.

The *Mar*-boats show their American heritage in their general beaminess, a trend that Fred Kohse encouraged in the design of his *Sleep-Robber* (1956). In the US the Alaska-limit seiner developed beam when the length was limited to a 58-foot maximum. In Canada, Sam Matsumoto was the pioneer of the beamier boat. He first demonstrated his belief in adding to the beam on seine boats when he designed the 48-foot *Eskimo* in 1940 at Prince Rupert. "I changed my father's ideas of narrow, long boats to a short, wide boat. Dad didn't agree with me, but this was the beginning of our modern type of seiner. The engines were getting smaller in size but they had more power [to push a wide boat through the water], so I made the engine room smaller and the fish hold larger. The *Eskimo* was the first one that I tried. The next was the halibut boat *Robert B* in 1941, and then I designed the *Universe*. When the war broke out, that boat was being finished off. I was sent to the road camp and Dad was allowed to stay and finish the boat."

The *Eskimo*, above, was the first boat that Sam Matsumoto built to take advantage of the new, smaller engines by adding beam. This was in 1940. His design ideas were interrupted by the war, but after the war he returned to the coast and built a yard at Dollarton on Burrard Inlet. (FT; AHB)

When Japanese Canadians were ejected from the West Coast, Sam was sent to build the houses for a camp at Alvida, BC, between McBride and Blue River. He had never built houses before but after boats they were easy. Toward the end of the war, Sam moved to Slocan and later to Nelson with his father and family. They began building rowboats for people in the area. Soon they were shipping their boats throughout the Kootenays, and into Washington, Idaho, the Okanagan and Waterton Lakes National Park in Alberta. But in an area where lakes are frozen for much of the year, boat sales were limited. As soon after the war as he was allowed, Sam travelled to Vancouver to check out the market for commercial fish boats. What he heard was encouraging. He returned to the Kootenays and built two 32-foot gillnetters which were shipped by flatcar to the coast. They both sold before they could be unloaded. One went directly to Great Slave Lake for fishing whitefish.

About 1949 Sam came down to the coast to look for a place to set up a shipyard. He decided he wanted to be on the Vancouver waterfront, so he bought the only available location, above the Second Narrows on the North Shore of Burrard Inlet at Dollarton. One of his first customers, Nelson Brothers, placed an order for three gillnetters, provided they could be built at the same cost as in Nelson's own yard. Fishermen came to look at the boats under construction and were so impressed that the Nelson Brothers' order was doubled to six boats and then to twelve. Orders flooded in from BC Packers, Canfisco, Cassiar Packing, and others. The Matsumoto yard began building twelve boats every six weeks, with a crew of about twelve workers, including Sam's dad and two brothers. The operation was so impressive that the same government that had declared Sam an "enemy alien" ten years earlier now funded the National Film Board to make a movie of this new chapter in the Matsumoto family history.

The design ideas that Sam Matsumoto had been working with before the war were still very much in his mind. In 1954 he designed and built two seine boats that caught the attention of everyone on the coast, not just for their beam but also for the dramatic flare to their bows. The *Zodiac Light* and the *Eva D II* carried the flamboyant styling of fifties-era consumer goods into commercial boat design. They may have owed some of their flare to a series of four boats built by the Canadian Fishing Company's Sterling Shipyard in 1949–50 to a design by the American naval architect H.C. Hanson. But the Canfisco boats had 17-foot beams on 60-foot registered lengths. The Matsumoto boats carried the same beam on only 53-foot registered lengths. The boats were well received and were followed by the slightly longer *Miss Georgina* and *New Venture* in 1955.

In 1955 Sam's prolific yard also built the *Ocean Star* and Bill Pitre's *Pacific Belle*. These were the first of a series of double-deckers. Many of Sam's customers were from Prince Rupert and knew his work from before the war. "After I started building these boats [a fisherman named] Erikson came over and said, 'I want a 72-footer.' That was the *Ocean Star*. Then Johnny Jonson came and I built one after the other. For Haugan I built the *Sunnfjord* and for Percy Pierce the *Blue Ocean*. The construction was really sound on all of these boats. The plank that is on the deck just where the stanchions come up to the bulwark is called the cover board. Most builders cut it out on three sides around the oak rib and have the sheer strake come up on the outside. Then they caulk around the stanchion. Well, you are bumped and it dries and gets loosened up and the oak starts to rot. We cut our frames right off at the sheer and the covering board came right over the sheer plank. Then the stanchions that came up for the bulwark were all made of gumwood.

"We also put a lock strake in the deck just in from the cover board.

Sam Matsumoto holds a picture of the *Northern Dawn* in his Dollarton office, in a photo taken shortly before he retired and sold the property. (AHB)

The *Miss Georgina*, built in 1954, continued the beamy style begun before the war with the *Eskimo* and the *Universe*. Note the 1950s streamlined influence in the sweep of the bulwark shear line, which is matched by the sweep of the dodger and the low house profile. (VMM)

Bill Pitre's *Pacific Belle* replaced his smaller boat of the same name. (CI)

This plank was a little thicker than the others and was cut down into the deck beam so that you can drive the caulking in hard and it won't spread the deck, it wouldn't push the planks over. It all locked tight. You don't get all that in other boats. The cost of building in those days was not so much. But of course, I never made money, I just built boats.

"The last boat was our *Northern Dawn*. It's a 76-footer and I built that on speculation in 1960. After that the price of herring went down flat. It took me six months to bring in the kiln-dried decking and the edge-grain lumber I did have. I couldn't go on building if I had to wait six months for materials. I sold that boat very cheap too. I used some aluminum around the rudder on the *Northern Dawn*. Pure aluminum had been used in boatbuilding many times since about 1890, rivetted because you weren't able to weld it. But it was just the pure aluminum, and salt water eroded the material.

"The United States and Japan were using alloys and gas welding. I made a trip to look into it. We made some aluminum gillnetters, and from there on it was aluminum and steel for our boats. We built the *Secord I* and the *Evening Star* in steel and then we built more seiners in aluminum."

The 1950s saw both the pinnacle and the end of wooden seine boat construction in BC. With the addition of beam and size, boats like the *Waldero* from Bensons Shipyard, the *Nanceda* from Menchions and the *Sleep-Robber* from Mercers became the object of most seine fishers' admiration. The postwar boom in the use of hydraulics encouraged the development of both the power block and the drum seine and greatly enhanced the efficiency of the seiners. Although there were a number of good seiners built in wood right into the 1970s, the preferred materials became welded steel, then aluminum and fibreglass. Metal or fibreglass construction saved on labour costs at the same time as high-quality woods were becoming harder to obtain. Then came the development of superior marine-grade aluminum alloys and plasma welding techniques. There was just no place left for wood in the marketplace.

From Russia With Love

The *Northern Dawn* was already four years old in 1965 when Madia Shebikhanova was born, halfway around the world in the Caucasus Mountains in the south of what was then the Soviet Union. When she was six, Madia moved with her mother thousands of miles east to the Kamchatka Peninsula on the Pacific. Madia grew up there with her stepfather who was trawlmaster on a Russian deep-sea trawler.

After completing secondary school, she entered the Petropavlovsk-Kamchatski Marine Academy, from which she graduated in 1990 as a navigator. She had already made a trip to the Bering Sea on a trawler in 1986 and now she went to sea on the Russian-Canadian joint venture hake fishery. But things were changing at home where people of her Sheshen nationality were being persecuted in the breakup of the Soviet Union. Things weren't much better on the ship where she was the junior of three navigation officers and one of the very few women in the Soviet fleet working at that level.

Along with several other members of the crew, Madia walked away from her ship while it was docked at Ballantyne pier on the Vancouver waterfront. After contacting the Russian consul to explain her reasons for leaving the vessel, she got work on a Canadian trawler and later met her partner, Sid Smith, the current owner-operator of the *Northern Dawn*. "I want to go salmon fishing with Sid," she said in the summer of 1993. "But he says there is only room for one skipper on the boat."

There is a twinkle in her eye as she explains that with two years of experience she could skipper a seiner. Her determination to date leaves little doubt that given the chance she will do just that.

Madia Shebikhanova and the *Northern Dawn*. (AHB)

THE DAVIDSON GIRL

I t rains a lot in Port Edward in January. When Nelson Brothers Fisheries had their herring reduction plant working there in the 1960s, the winter herring fleet would come in from Hecate Strait through the misting rain to moor alongside the empty gillnet racks on the floats below the plant. There they waited their turn to move under the pump that lifted their night's catch up to the conveyor to send the fish back to the big storage bins. The water of the harbour was glass frosted by the drizzle. Crews moved slowly between boats, feeling the tautness of tired muscles that is the reward of long hours of hard work on heaving decks. As they traded *Playboy* magazines and stories of last night's fishing, the smell of bacon drifted up from the galley, defining a small circle of pork smell in the midst of the heavy, sweet presence of fish.

There were seventy-one boats fishing herring in the winter of 1961. The biggest and best of the British Columbia seine fleet, they were mostly double-deckers like the *Western Producer*, *Nanceda*, *New Queen* and *Belina*. A few of the older, single-decked pilchard boats like the *Ribac* and *San Jose*, second-class company boats assigned to Indian skippers, lay loaded at the dock with inches of water covering their decks.

Nearby, a half-dozen smaller boats clustered against the tall pilings. No fire burned in their galleys. Inverted buckets protected their exhaust pipes from the steady rain that ran off the decks, leaving trails of green algae on the white cabin sides and merging with the Nelson Brothers' green-painted hulls. The windows stared emptily from brown painted frames. Everything about the vessels said: "company salmon boats." Too small to fish herring, they were painted by wage-earning shipyard workers rather than boat-proud owners. But the flare of the bulwarks, the rake to the shear line, and the solid stance with which they sat in the water suggested an eclipsed elegance. These had not always been company boats.

"Those," explained one of the herring crew, "are the Haida boats. They built a whole bunch of them about ten years ago but the company took them all away."

Charlie Clarke, skipper of the *Western Producer* and Ritchie Nelson's head skipper, told the story. "Ritchie Nelson wanted the fish from the Queen Charlotte Islands and he knew those Natives up at Massett could catch them so he gave them money toward those boats. They built them shallow to go into Massett Inlet. I heard they were so shallow that you could stand in the hatch and pee overboard, but I think they are really deeper than that. Anyway, those guys didn't fish hard enough to pay them off. So Ritchie had to take them away."

THE *DAVIDSON GIRL*

Haida elder Florence Davidson remembers a different version of the story. She was there when her family built the *Davidson Girl* in 1952, but her knowledge of Haida history extends much further back. The story of the seine boats can only be understood in that context.

"We were on the island before the Flood. Some people were saved on big canoes. There's something on the hill for TV. That's where the big canoes anchored up. When they looked in the water they used to see all kinds of houses and poles in front of the houses below. That's why, when they got on the dry land, they started making houses and put designs at the front and poles at the front. . . .

"When I was little, Andrew Brown built a boat with Daniel Stanley. That's why they called him Captain Brown. That was about 1907 or 1908 when they built that boat.

"Douglas Edenshaw built the *Laverne Marie. Chief Weah* was Mathew Yeoman's boat, he built this boat in 1927 way before the Nelson Brothers came to the islands. Henry White built the *Haida Girl* in 1931. . . . Andrew York built the *Massett Maid* for Ernie Yeltatzie in 1948. He built the

The *Davidson Girl* runs down Prince Rupert harbour. On board are Robert Davidson Sr. with his sons Claude and Reg, together with young Ernie Colinson. (JW/FT)

Geoffrey White at the wheel of the *Haida Warrior* in the early 1950s at Prince Rupert. Her white paint job proudly signifies a privately owned vessel, but this did not last long. (JW/FT)

Haida Brave the next year for himself. I think the *Laverne Marie* was built for Nelson Brothers in 1952.

"They built the *Davidson Girl* right in front of our house. . . . A young man came to me, he called me Sister because his uncle adopted me for his niece. My late husband was married to that family. He said, 'Sister, don't trust white man. Any time they want they'll snatch your boat away. Don't buy too much for it too. You know the businessman, any time they want to take it away they'll snatch it away like nothing.'

"So I told Robert, my husband, 'You get a big notebook and our son Claude will write everything. Every cent you spend on the boat, you make him write it down and your time working on the boat, write that down too.' They launched it in 1951 but it was too late to fish that year. My husband bought lots of expensive things to put on the boat.

"So Geoffrey White almost pay up his boat the *Haida Warrior*, just a few thousand more to pay, and the *Adelaide J*, George Jones built that boat way before Nelson Brothers came, but even that one, they took all the boats away from them."

The Haida people were not willing to give up their boats without question. A number of them travelled to Prince Rupert to meet with Ritchie Nelson. "We were all ready to go to Port Edward for the salmon and we got the news. I guess they sent a telegram. That's when they had the court about it. But my people think it's hard for them."

Florence told her husband: "'You take the book with you, Claude will read it out for you.' Mr. Nelson didn't know they had the book. 'Bob,' he said to my husband, 'you got everything written down.'

"'Yes, Claude will let you hear everything. We used to work, he wrote that down too.' Mr. Nelson, it look just like he flop down in the chair.

"We lost our home in a fire when the boat was being built. . . . When Nelson heard everything that we bought for the boat and when they worked on it with the hours and the day, he sat for a while and said, 'We'll give you so much money so you'll build another house.' No one else got any money from Nelsons, just us. . . . ."

The boats were all lost to the Haida. The British court recognized the value of the Nelson Brothers' invested capital over the Haida investment of labour. The system of "grubstaking" potential producers so that they would go out and trap enough fur or catch enough fish to meet their own needs and

After years with her hull painted green and her name changed to *Western Wave*, the *Haida Warrior* was repainted, renamed and returned to her Haida Gwaii home in time to take part in the 1985 blockade at Lyell Island with the *Haida Raider* (ex-*Ed II*) and the *Haida Spirit* (ex-*Otard*).
(*Vancouver Sun*)

return profit to the investor had worked well enough in the days of the Hudson's Bay Company, but the Haida were not willing to harvest the salmon as aggressively as the Euro-Canadian society demanded.

Florence's son Claude, father of well-known Haida artist Robert Davidson Jr., continued to fish in a small troller. But when his wife Vivian was lost overboard he sold the boat in despair, only to find later that he had sold his licence and had become a Haida whose right to sell fish in his own land was limited by a non-Haida government thousands of miles to the east.

In recent years, actions on the part of the Haida people have done much to reverse this trend. A number of the old boats that were lost in the 1950s have been bought back by the Haida. When the *Haida Warrior* was lost to the company, she was renamed the *Western Wave*. She is back now under her original name. So is the *Haida Brave*. The *Gwen Rose* has been bought by Willis Crosby and renamed *Tanu Warrior* after one of the old villages.

In addition to these classic wooden boats, modern aluminum seine boats have been added to the Haida fleet. The aboriginal claims of most First Nations groups include a significant share of the commercial fishing resource. The old herring reduction plant and cannery at Port Edward was torn down some years ago, but in the Haida Gwaii the fish still return to the spawning rivers and beaches and the Haida fish them with their seine boats.

"FERGY" FERGUSON

Fergy's first boat was the wooden baby drum seiner *Misty Lady II*, built in 1955 at a time when drum seining was just developing. (Ferguson family)

W hen the first wooden version of the *Vampy* showed up on the fishing grounds in the early 1960s, she raised a lot of eyebrows with her innovative cabin design. Today, cabins of the same or similar design are standard on most seine boats. O.C. "Fergy" Ferguson, the *Vampy*'s skipper, was also her builder and designer—continuing a tradition of boatbuilding fishers in the BC fishing fleet. But that first *Vampy* seine boat was not the first boat that Fergy built, nor was it the first boat that he fished.

Fergy grew up in Vancouver's Dunbar district. As a youngster during the Hungry Thirties he helped his brothers fish the north arm of the Fraser in a row skiff to make a few extra dollars for the family. Then, during the war, he and a friend bought an old clunker of a gillnetter that was tied up under the Burrard Street Bridge and took it to Rivers Inlet where they pulled net by hand.

After the war Fergy started building pleasure boats. His first *Vampy* was a runabout with a V-8 inboard. He put a net on the back and fished Canoe Pass, the southernmost arm of the Fraser estuary. The first bona fide fish boat that he built was a 27-foot gillnetter, the *Vampy III*, followed by another, slightly longer gillnetter. By the 1950s, Fergy had made up his mind that he should have a drum seiner, but he couldn't afford it. So he

designed and built the little *Misty Lady II* in 1955 and fished her at Sheringham and in Johnstone Strait for four years. That boat made a lot of money for a lot of people. She was owned by Sandy Buntin for several years and most recently by Bill Gladstone up in Bella Bella. At 42 feet, with a 14-foot beam, she is about the minimum size for a drum seiner.

From his experience with the *Misty Lady II*, Fergy built the original wooden *Vampy* seiner. He wanted a double-decker style cabin on a boat large enough to fish the West Coast but still suitable for Johnstone Strait and drum seining. The *Vampy* was 54 feet long with a 17-foot beam. The crew slept on the main deck level, which was a dramatic break with tradition for such a small boat and a far cry from the *Misty Lady II* with her four men in the fo'c'sle. From the galley, four steps led up to the wheelhouse and the skipper's bunk.

The *Vampy* was the first boat with a pursing winch and rod rigging and Swann made a geared drum drive in place of the automotive drive that some of the other boats were using. Another innovation was that Fergy was able to do the pursing, the spooling and the drumming from a set of controls mounted by the mast on top of the cabin. The drum was in a well, so the spoolers were easily visible. He used the same purse line in the pursing winch for four years, unheard of in those days when winch heads had little water lines to cool the new nylon lines.

The first *Vampy* raised eyebrows with her split-level cabin arrangement, which gave the effect of a little double-decker. (Ferguson family)

In the mid-1960s Fergy built the wooden *Vampy II*. This time the design was more conventional. He did not plan to fish the boat himself and he didn't want another skipper faulting the boat if he wasn't catching any fish. The original *Vampy* was later lost in Hecate Strait, but *Vampy II* is still going under a different name.

After selling the *Vampy II*, Fergy kept the name and built a 60-foot aluminum version. This time he went all out with experimentation. He installed two V-8 Jimmys and tried to get speed. "Every boat is a compromise," Fergy says. "I thought that it should be long and narrow but to work a semi-planing hull I should have had a little more beam to carry the weight." In addition to the innovation of twin engines, the new Vampy II had the drum set right behind the cabin. The bag with the fish could be pulled up the ramp and spilled right into the hold as is done on the big draggers. The dead (rowed) skiff was set right off the ramp. The power take-off was driven by both engines, and the boat had aluminum piping for the hydraulics. Another first was an aluminum-lined fish hold with a sandwich of insulation between the hull plates and the lining. In practice this didn't work very well because the aluminum lining had to be fastened to the frames, causing a loss of insulation efficiency.

Fergy and one welder built the entire boat in six months. Only a few weeks after he launched it, in 1968, he was approached by the United Nations Food and Agricultural Organization to take the *Vampy II* to Venezuela to teach seining. Once that contract was complete, he sold the boat in Venezuela. But it didn't take long for Fergy to build a replacement, another aluminum model, only this time with one engine, a 3408 Cat. Named the *Iffl Biffl* after the kind of bureaucrats who made Fergy impatient, she has since been sold to Dickie Michelson and renamed the *Island Breeze*.

Fergy was thinking to retire, but then somebody showed up wanting a design for a 75-foot steel boat. He did a design for the *Leader IV* and then in 1978 built the *Vampy I* to a similar design. The major innovation in these boats was that they had the engine in the stern. Fergy thought it was ridiculous to have an engine in the bow and a propeller in the stern with 40 feet of shaft and bearings in between. The *Vampy I* also has bow and stern thrusters, a quadrapod mast, and twin exhausts which were incorporated in the mounting for the drum.

In 1987 Fergy embarked on a matched set of new steel boats, one for himself and the other for Mario Carr. This design featured a stern which leaves the water much like a duck. The hull sweeps up at the stern so that the vertical transom starts just at the water line. The idea was to

**Fergy's aluminum *Vampy II* featured a ramp stern, twin propellers and a drum mounted ahead of the hatch.
(Ferguson family)**

"FERGY" FERGUSON

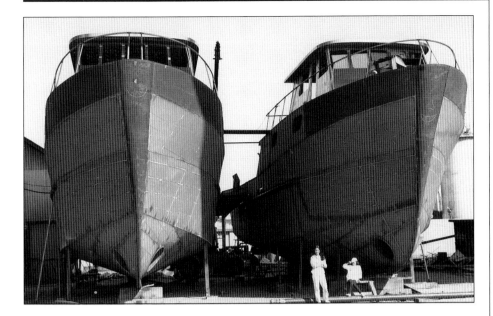

When Fergy retired he "got tired of chasing that little white ball around." So he built two more seine boats in a rented space under the bridge to the Vancouver Airport. One was his own *Krista Gail*. (AHB)

provide a good entrance in the water and a broad stern to carry the weight of the net and gear, yet leave the water clean with no tumbling action. His own boat, the *Krista Gail*, is powered by a 350-hp Penta 121, while the other, the *Pacific Venture I*, has a 3408 Cat generating 435 horses.

As he contemplated his latest job, Fergy summed up his philosophy as a boatbuilder, one that has kept him doing it for fifty years. "There is nothing more satisfying or challenging than building a boat. A boat is a compromise of so many variations. There are a million ways of doing it, especially when you start with a blank piece of paper."

Aubrey Roberts, on the *Western Investor*, prepares to lower his herring pump into the water as his brother Tony comes on his cork line with the *Western King*. They are fishing the Barkley Sound herring opening in 1989. (BG)

A HARD ROW TO HOE

"**A** guy phoned you from the Deep Sea Trollers Association," said my wife. "He sure seemed cranky about
something."

All I could do was laugh. I had been expect-
ing a call from the executive director of the Deep Sea
Trawlers Association. My wife was raised on an interior ranch and wasn't
too clear on the difference between boats that fish salmon by trolling a
hook and line and boats that trawl a big bag of a net along the bottom or
through the mid-water depths in search of cod, flounder, pollock and
hake. Trawl fishers don't like being called trollers. And just to keep it con-
fusing, they don't usually call their boats trawlers. Their boats are called
draggers, except when you are getting sort of official. By this time my wife
had lost interest.

Trawlers are a breed apart from other West Coast fishers. Salmon
net-fishers, who fish the limited one- or two-day summer openings, shake
their heads at the driven pace of the draggers coming and going from the
grounds to deliver at Prince Rupert, Port Hardy, Ucluelet or Vancouver, in
a seemingly unending routine. It is a hard and wearing life. Draggers of
over 100 feet have disapeared without a trace. Their loads can shift in an
instant; a few minutes of downflooding into the hull through an open

**Hank McBride has fished lots of less
comfortable boats, but the hours spent
surrounded by electronic sonars,
sounders and plotters while looking for
fish can still get tedious. (AHB)**

The *Gail Bernice* carries two nets on her stern drums. The "doors" hanging on the stern quarters are designed to keep the mouth of the net spread open. (AHB)

door or hatch and a big steel boat goes down like a stone.

These are not the things that Captain Hank McBride likes to talk about during the long hours he spends in the wheelhouse of his 96-foot dragger, the *Gail Bernice*, especially when he is on fish.

It is 10:30 on a mid-November morning in the Gulf of Georgia. Splashes of yellow-trimmed red move across the screen of the colour sounder at the 60-fathom mark. Each splash represents a small school of pollock, with smaller schools and individual fish showing as bright yellow stars against the dark sky of the video screen. By the time the colours have progressed to the left of the screen, there are equivalent marks showing on the paper printout from the net sounder. Here the fish show as grey scratches between two firm lines marking the head rope and the foot rope that form the top and bottom of the midwater trawl's mouth.

The gear has been in the water since 8:00 a.m., after a half hour of searching had shown schools of pollock forming with the late-breaking November daylight. The main schools have been showing between 40 and 50 fathoms, so Hank had ordered the net let out to the sixth mark on the cables. With 25 fathoms between marks, that would put the net 150 fathoms from the stern of the boat, at a depth of 50 fathoms. By 9:15, there were enough fish in the net for Hank to order it hauled back one mark, to bring it up 10 vertical fathoms to the original depth, as the fish cause it to sink. The mouth of the net is about 15 fathoms deep between the head and foot ropes.

According to the digital readout on the top of the video plotter, we are making 2.7 knots. The 940-hp Cummins main engine is turning at 1600 rpm. There is a 20-mph southeaster driving a cold rain against the wheelhouse windows and the big boat lifts easily as it comes into the wind and makes the turn, preparing to drag back over the same water that we have just come through. The course shows on the video plotter as a white line connecting way markers that Hank has saved in the machine's computerized memory, from successful tows at this very spot the previous winter. But now fishing is spotty, and Hank comments to a deckhand, "It's a long row to hoe, but we'll hoe it to the end."

This comment stirs the skipper's memory of growing up on his dad's farm on the Serpentine River near White Rock, and the half-mile rows of corn that the young boy had to hoe while others went off to July First celebrations. His father had encouraged him with the same comment. Remembering just how tough things were for farmers and fishers in the Dirty Thirties, Hank tells the story of the American skipper-owner of an 80-foot halibut schooner who came up the river and tied at the farm's dock down by the barn. They loaded a mixed cargo of booze and hay onto the boat, with the whiskey well hidden under the hay to escape detection by the American revenue agents. When Hank's dad remarked on the chance the skipper was taking, the old man responded, "Ya! Well, it's no chance. If I don't make this trip, the bank will take my boat next week for sure."

11:00 a.m. Hank turns the dial on the auto pilot to bring the boat in closer to the shallowing edge near the steep shoreline. Five minutes later he sees the bottom coming up quickly toward the keel of the boat. It comes straight up to 10 fathoms as Hank hits the button to start the auxiliary engine that runs the hydraulic winches. He calls the crew to take up on the starboard cable to help take the boat out from the menace of a major snag, then spins the dial on the auto pilot and speeds the main engine. The bow of the boat swings out and the gear, far astern of the boat, follows suit just in time to avoid a collision with the bottom. By 11:12, the gear is clear of the bottom and is catching fish right along the edge. Hank settles back in the worn skipper's chair.

Noon. Fish show on the colour sounder under the boat at 40 fathoms. Hank speeds up, to sail the 500-foot rope trawl with its big steel doors up 10 fathoms to catch the fish. We've been on this tow for four hours now and the three crew are having another cup of coffee in the galley, telling stories of cod ends not tied properly and the things that skippers say when the net is brought up after a four-hour tow to find an open cod end and no fish. We all have our nightmares.

Deckhand Alan Hodgson standing by to let out one of the two cables that tow the net. (AHB)

The midwater trawl net is set from the starboard drum. (AHB)

1:00 p.m. The *Gail Bernice* arrives back at the point where the trawl was set out five hours before. Hank calls the three-man crew to haul the net back.

1:15 p.m. The cod end of the net is floating on the surface like a fat fish sausage. The boat doesn't have a stern ramp so the net is brought around to the starboard side and a smaller bag of fish is split off. This bag is lifted aboard and lowered over an open hatch cover. The knot in the cod end is good but a little too tight, so deckhand Alan Hodgson has to hook the end to a little winch to trip the knot and spill the fish into the hold. The end is refastened and a second and then a third split of fish are brought aboard.

1:45 p.m. The gear is back in the water and the afternoon tow begins. Hank estimates 16,000 pounds of pollock. At 11 cents US per pound, it will pay the fuel for the trip, which has not been productive up to this point. But then November dragging in the Gulf of Georgia is considered a pretty hit-and-miss kind of fishing. On the previous trip the *Gail Bernice* had spent two days cruising without seeing enough fish to bother setting the gear. Then a solid bar of red showed on the sounder where nothing had been earlier. In two tows, one of twenty minutes and the other of forty minutes, they took over 100 tons of hake, enough to load the *Gail Bernice*, with her 240,000 pound capacity.

Hank admits to getting a little discouraged when fishing is slow, but in fifty years of hoeing the row and working the tow, he has learned the slow times are as much a part of the business as the good times. He started gillnetting with his dad when he was only nine. His first boat was a ten-foot skiff. In 1946, when he was nineteen, he and his dad had the *Sharon M* built at Mercers Star Shipyards in New Westminster. "There was no BS about their boats," remembers Hank. "They even had the eight cup hooks in the contract."

The *Sharon M* was designed by Gerald Seaton, who was supervisor at BC Packers' Celtic Shipyard. She is related to the *Active Pass*, the *Louisa Todd* and several other boats. The first year they went halibut fishing and dragging with the new boat. They hired John "strong-as-a-gearmatic-winch" Johanson as skipper, and the crew were "Silent Gilbert," "Nine Fingers John," "Eino the Finn" and "Old Billy," who worked as a fisher into his eighties.

5:05 p.m. Hank has ordered the net hauled back, the doors are up in the stern davits and the net is being wound onto the starboard drum. The port-side drum holds a net that is virtually identical to the one being fished, but it is 25 percent smaller. It is used as a spare, or for fishing difficult ground where greater manoeuvrability is needed. The cod end on the net comes up with about a third of the catch the first tow produced, but this was only a three-hour tow.

A "split" of pollock is hoisted aboard. When this photo was taken in 1988 it was necessary to split the catch into manageable amounts before lifting it aboard. The boat has since had a ramp stern added, making the recovery of the net with its catch much easier. (AHB)

It's dark and the fishing isn't so good after dark. There were a few small dogfish mixed in the last tow, and there would be more after dark, so Hank finds a quiet bay with a log boom to spend the night. With one more day of fishing he expects to get the balance of the 50,000 pounds of pollock that the plant in Anacortes, Washington has requested. Then it will be time to head back to the upper Gulf to look for some hake to top off the load. Although he has already spent two days searching for hake before getting the order for the pollock, it will be hit-or-miss. "It's time to tie her up after this trip and get the boat overhauled," Hank says. He plans to put one of the new-style kort nozzles around the prop to increase the thrust and towing power.

In the galley, Alan Hodgson is poring over back issues of the *Westcoast Fisherman* looking for an affordable C-licence boat to fish cod. Randy Taylor, who is acting engineer this trip, is worrying over a stripped wheel on one of the winch brakes. Phil Gotch is cooking up enough supper to feed a crew twice the size, but most of it will be eaten at the first sitting and more will be snacked on later in the evening while the men watch some of their dwindling supply of videos. Phil has fished on a lot of boats since he started gillnet fishing with a skiff and 50 fathoms of net. He has been with Hank for fourteen years, except for a year with Charlie Drake on the *Western Ocean*. He was with Hank when his last *Gail Bernice* burned off the West Coast in 1981.

The last *Gail Bernice* was a 130-foot steel boat built in Holland in 1956. Hank had traded in his first *Gail Bernice* (now the *Quadra Isle*) for her in 1969. He had her built at Ben-son's in 1957. He dragged with her through the years when ground fish had little market value. "Those were tough years," he says. "I fished mink food. There were lots of fish, but no price. We would put in 180,000 pounds a trip at 2 cents a pound. You got the fish so fast you only used twelve or fourteen tons of ice."

The third and latest *Gail Bernice* is also a Benson's hull. She was built in 1981 for Dick Sims as the *Simstar*. With her 27-foot beam on a length of 95 feet, it is an under-statement to call her comfortable. Her size and the sophistication of her RSW system and her elec-tronics—in addition to radios, she has a big paper sounder, colour video sounder, net sounder, auto pilot, loran, plotter and two radars—show just how much the BC trawl fishery has changed since the demand for ground fish has grown. Hank says that Wes Johnson, now a Fisheries consultant in Washington, is the grand-father of mid-water trawling in BC. "He began researching it in the 1950s. At that time, even without net sounders, we could get 200 tons of herring a week down around Swanson Channel."

Wes Johnson may be the grandfather of mid-water trawling, but by all the accounts of his fellow fishermen, Hank McBride is one of the fathers of this trawl method. He has been at it as long and as hard as any-one on the coast and there is no doubt that as he finishes hoeing each long row, he isn't going to waste much time starting on the next one.

A few good hours on deck working gear in the wind and rain are rewarded by cook Phil Gotch's work in the galley. Food and videos help pass the time for the modern fisher, who can still get up a pretty mean game of crib or poker. (AHB)

THE ALUMINUM AGE

As the cost of labour increased and people's willingness to spend years learning the art of wooden boatbuilding decreased through the 1960s, fewer and fewer of the bigger wooden boats were built. Steel had gained considerable acceptance for boats in the 70-foot range as efficient welding techniques brought the cost down from the old method of building hulls of riveted plates. But in the end it was aluminum that gained the place of dominance for seine boats in the BC fishing industry.

This was not true elsewhere. One morning in the late 1980s I sat on the gunwale of a little 48-foot steel dragger in the south England fishing port of Weymouth, as the skipper and I compared the tribulations of fishers in England and Canada. He told me that the Common Market was giving too much of the cod quota to France, and I told him about the Canada–US salmon treaty. He explained his trawl gear, and I showed him a photograph of a fine new BC aluminum seiner. He looked at the picture and handed it back. "Ally, ay," he said with a note of disdain in his voice. "Tried that here. Didn't work!"

I had heard similar remarks in other places. They were common in the galley talk on the West Coast in the early 1970s, and at the time I had believed them. "Aluminum is too light so it floats too high and rolls over easily," was one of these ideas. The most damning, and the closest to the truth, was that other metals on the boat would cause electrolysis which would "eat up" the boat as it sat in the water. Those of us who loved the romance of wooden boats and enjoyed the conservative rhetoric of the fish deck, decried the first aluminum boats long and hard. But others, with their eye on lower maintenance costs, appreciated that unlike wooden boats, the aluminum never needed recaulking, and unlike both wood and steel, aluminum never had to be painted.

My appreciation of aluminum was delayed for over a decade until I had the opportunity to meet Al Renke. Born in Germany, Al grew up during the war. He began his apprenticeship in a German foundry before coming to Canada in the 1950s. On his first job in a Canadian foundry, he amazed the supervisors with the quality of his work. But as Al explained years later, this was simply the level expected in the demanding German system. It was this attention to detail that he applied to learning to cut, shape and weld aluminum. At first he had a small shop doing general fabrication, but a job to build some fish-cutting tables led to a contract to build a small aluminum gillnetter. It was Al's first boat.

In 1972, Al moved to a location on the Fraser River dyke near the

Al Renke, with his European training, set the standard for aluminum fabrication. (AHB)

THE ALUMINUM AGE

The design for a typical Shore seiner as drawn by Grant Brandlmayr. This boat is 68 feet over all and has four fish hold compartments with a walkway from the engine room to the lazarette.

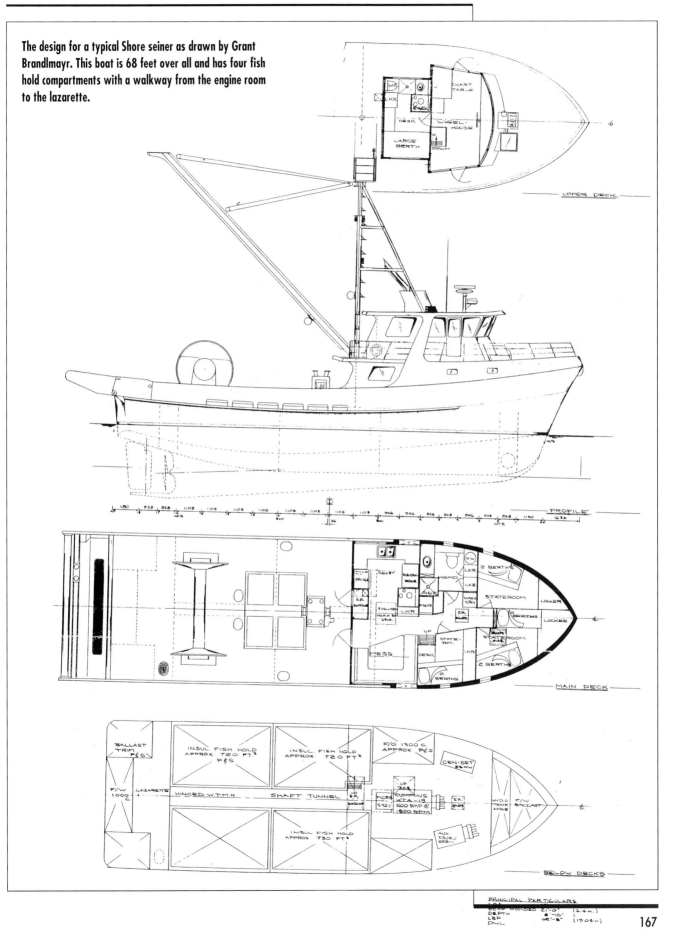

Bill Wilson's seiner *Kynoc* under construction in 1986. (AHB)

BC Packers' Queensborough Shipyard. Shortly after moving he got the first contract to build an aluminum seiner, for Ross Aleksich. Named the *Westfield* (since renamed *Progressor*), it was a 64-footer, designed by Al McIlwain. This was probably the first seiner with a bow thruster. The small propeller mounted in a tunnel located crosswise in the bow of the boat allows the skipper to push the bow of the boat around while fishing. The innovation caught on with the fleet almost immediately as it made it possible for a boat to avoid being entangled in its own net. Seine crews would no longer have to strain their backs pulling in the cork line in order to pull the web back around to the right side of the boat.

The young company's next order came from John Polonio for the 70-foot *Pacific Quest*. With a 20-foot beam, she impressed the industry with the amount of boat that could be fitted into that length. The salmon seine fishery was about to become a limited-entry fishery with about 520 licensed salmon seiners. New boats were only possible if an old boat was retired from the fishery. Because the new boat had to be built to the retired boat's licensed length, a couple of feet were inevitably added to the beam. Additional freeboard and fish-packing capacity were added by using a hard chine design. This is a squaring of the point where the boat's side turns toward the keel. On the old wooden boat this was a graceful and rounded line; now the beam of the boat continued down from the deck level to be about 80 percent of the deck beam on the first aluminum boats. On later boats the deck beam was increased and the percentage of chine beam was

THE ALUMINUM AGE

Gerry Dobrilla's seiner *Pacific Discovery* nearly ready for launching. The black material on the stern quarter is nylon to protect the aluminum from the banging of the trawl doors. The extension out from the stern at deck level is for a hydraulically operated tilt stern for bringing fish aboard. (AHB)

also increased. The old wooden hull had evolved from sailboat designs which required a hull form that would move easily through the water. The new hard-chined hulls were made possible by the smaller and lighter, but increasingly powerful, turbo-charged marine diesel engines.

The demand for Al's boats grew so rapidly that naval architect David Moore came to work full time at Shore Boats. There he designed the *Pacific Quest*. A lot of the earlier aluminum boats by other builders were patchwork in appearance. Some builders ground their welds, which smoothed or faired the hull. It also interfered with the natural oxidation process, which seals the surface of aluminum plate, and added to the patchwork look. At Shore, Al stressed making a weld look nice the first time. To make a design that faired the hull, he developed a pattern of frames so that a large sheet of aluminum was wrapped onto the hull. This allowed the use of larger sheets with fewer welds. Because every sheet had a slightly convex exterior, they also had greater strength. Most important, they created the rounded lines that made a Shore seiner look like a boat rather than a baking pan.

Over the next two decades Shore Boats built seiners that replaced more than a tenth of the BC salmon seine fleet. When David Moore retired, Grant Brandlmayr, son of the well-known yacht designer John Brandlmayr, assumed the lead design role. A representative pair of boats, built in 1986, were the 56-foot long by 19-foot wide *Katrena Leslie I* for Alfred "Hutch" Hunt, back for his second Shore boat, and the 60- by 20-foot *Kynoc* built for Bill Wilson of Bella Bella.

The *Kynoc*, named for the serene inlet and river just north and inland from Bella Bella, opened a major generation gap with the little wooden seiner whose licence she was built to. The raised fo'c'sle design has evolved on all modern seiners. This has the effect of carrying the galley and sleeping area right to the sides of the boat, giving twice the living space. On a seiner of this type, the galley is usually on one side at the rear of the cabin with a dining booth built restaurant-style into the other side. Just ahead of the galley, a head with shower is on one side and the skipper's cabin on the other side. A ladder or stairway leads up to the wheelhouse, and a couple of stairs take one down into the crew quarters in the

Bill Wilson on the *Kynoc*. (AHB)

forward part of the fo'c'sle. Smaller crews, made possible by the hydraulic powered net drum, bow thruster, pursing winch and tilt stern, mean fewer bunks in the crew quarters. Most modern boats operate with a crew of five compared to the old crew of eight when nets were pulled by hand. On the old boats the crew climbed down into the engine room before going forward to the cramped fo'c'sle, where they suffered through hot summer nights in the heat from the big old diesel engine and the smell of diesel fuel coming up from the bilges.

On the *Kynoc* a ladder leads down to the engine room where hydraulic pumps and electric generators are run by auxiliary engines with as much power as the main engine in a narrow wooden boat. The main engine on the newer boat produces 425-hp, enough to push the wide, boxy hull through the water as fast as the mysterious physics of displacement hull speed will allow. Where an old 60-foot wooden seiner would have had a chain-and-cable steering system turning a quadrant on the rudder post, the new boat has a full power hydraulic steering system. The wheel in the wheelhouse is largely decorative. The boat is steered by auto pilot which can be overruled by the turn of a button on a remote box held in the hand of the person in the pilot's chair.

The *Kynoc* brings a bag of fish aboard with the aid of her tilt stern. The bow is cut back at the deck level to save a few inches of length for licensing purposes. (AHB)

The old wooden boat of this size would have one large open fish hold with provision for "pen" boards to be dropped into slots to "pen off" areas for sorting or icing fish. New boats have refrigerated brine tanks, or slush systems, that can be used to keep fish just short of frozen so that they can be held for the short one- or two-day openings and taken right to the plant rather than delivered to a packer. The *Kynoc* is designed with four separate holds, each of which is lined with insulating foam and fibre-glassed. Piping allows pumps to circulate the cooling water. These are managed from the engine room where a tunnel passes between the two port and two starboard tanks to allow access directly back to the lazarette.

The *Kynoc* has fished well for Bill Wilson on both salmon and herring. He has added a few refinements over the years. A government-funded experiment added a bulbous bow like those on the big freighters to see if it would improve the fuel economy. Bill has also experimented with on-board ice-makers in a never-ending quest to maintain the limited number of fish available in the best condition possible. Crew comfort and fishing efficiency aside, the most important thing about the new boats is the manner in which they maintain the quality of the fish from the time that they enter the net to the point at which they are pumped from the seiner's hold. The boats may not have the same panache as the old-timers, but the salmon that come out of their holds are of incomparably better quality.

Engineer Kim Anderson on the *Kynoc* checks out the complex of hydraulic lines that operate the drum, the spooler and the tilt stern. (AHB)

Of Boats and Bluffs

There is a classic story on the coast that has many versions, but a common ending. It starts with a long, hard weekend in port or a long, hard week of fishing. Either way the crew are tired and the person at the wheel falls asleep. The old boat wanders around for a while and then, full speed ahead, crashes into a sheer rock bluff.

The effect is explosive. As the bow stem hits the solid mass of Vancouver Island, or some such immovable object, the boat keeps moving for three or four feet. This has a dramatic influence on the shape of the bow of a heavy wooden boat: the plank ends splinter and wrap themselves around the bow stem. If the boat is travelling fast enough the bow will generally be opened up enough to allow the sea ample access to the fo'c'sle, so the boat will fill and sink. If the drop-off is right under the bluff or if a good sea is running, the vessel may well be lost. With luck the boat will settle in the shallow, where a patch can be applied. From there it is towed to town for the long job of having its bow rebuilt, while the engine is pulled and overhauled and the thousands of other jobs that have to be done to a boat that has sunk are taken care of.

On August 13, 1987 a new version of this story joined the sagas of Pacific boats and bluffs. It doesn't start with a hard weekend or even hard fishing. In fact, no one fell asleep at the wheel. The story is strictly high-tech from cause to result on Alfred "Hutch" Hunt's boat, the *Katrena Leslie*. On that fateful day the well-known Fort Rupert fisher turned the bow of his year-old Shore-built aluminum seiner south toward her home waters in Johnstone Strait. He and his crew had been "up North" all summer and the clear green waters of the strait, with the promise of some Chilko sockeye, were especially appealing.

Hutch says they headed for that sockeye via Grenville Channel, a straight and narrow marine course sometimes called Granville Street after that long, straight thoroughfare in Vancouver. About 45 miles long, the channel is a spectacular slash through coastal rock with steep-sided shores both above and below the surface of the sheltered waters. At its narrowest it is as thin as

"Hutch" Hunt. (AHB)

165 feet wide. Throughout its length, there is ample room for boats to pass, so it makes for an easy wheel watch. But even here Hutch maintains the common practice of two people on the wheel when travelling at night. "I had it on auto pilot," he explains. "You just sit there and flick the spotlight on once in a while for logs."

When Hutch went to his bunk for a little rest, he had the engineer take over the wheel. The engineer sat down in the pilot chair, turned off the auto pilot and steered manually. The second man on watch came up into the wheelhouse a short time later and the engineer went below to grease the shaft bearings. The deckhand assumed the boat was still on auto pilot, and because the channel is so straight and narrow, there would be no compass course to follow and no need to give the course to the new person when passing off the wheel.

The *Katrena Leslie* was travelling light and making close to 11 knots. The man in the wheelhouse turned to make an entry in the log book, while down below the engineer walked back from the engine room into the narrow shaft tunnel between the fish holds. The boat's bow entered a tide rip, and instantly it swung toward the near shore.

"We hit the bluff," Hutch remembers. "Full speed, about forty minutes after I got off the wheel. I had been sleeping and I slid ahead and hit the wall. My engineer came flying out of the tunnel and hit the back of the engine. He cracked four ribs and his lung collapsed a bit, but he is OK. My wife hurt her back and it still bothers her.

"When I ran up to the wheelhouse and threw it out of gear, there were trees right in front of the window. There was no rock, no beach, nothing—just the bluff. The poor engineer came crawling out of the engine room and I sent one of the guys that was in good shape down to check it all over. I told him to set there for a while and see if there was any water coming in. We just drifted off the beach. We checked everything and it was fine, so we just took off right to Shearwater."

The *Katrena Leslie* made a normal fourteen-hour run to Shearwater just across from the community of Bella Bella, and the boat was hauled out on the marine ways there. The bow had hit the rock at the curve below the waterline, the part of the bow that would be called the forefoot on a wooden boat. A perpendicular hung from the top of the damaged section would show about 11 inches bent inward from hitting the rock. That means that the 60-ton weight of the boat and gear, travelling at 11 knots, were stopped in less than a foot. This section of the bow below the waterline has a fresh water tank built into it that is used to adjust the vessel's trim and as a reserve fresh water supply. But its prime purpose is to serve as a collision bulkhead for just such occurrences. "There was just a little three-inch crack in there," says Hutch. "When we got up on the ways there wasn't even much water running out of there. Even if we hadn't had the tank in the bow, I

Al Renke compares the damaged bow piece with the repaired area.

don't think we would have had any problem."

With fresh water tanks in the stern, the fish hold, the fuel tanks and the forward tank, there is really only a very short piece of the hull, between the fuel tanks and the forward fish tanks, that is not effectively double hulled.

At Shearwater the rough pieces were ground off and the cracks were welded, about six hours' work. A careful check was made of the engine alignment and any other part of the boat that could have been damaged by the impact, but no problems were found. Hutch went on to the fishing grounds.

It wasn't until the end of September that it was convenient to put the boat up at the Queensborough Shipyards and cut away the damaged portion of the bow and rebuild it. The job took four working days and the boat was ready for the next opening.

Hutch has skippered boats since 1949, and he has owned some fine wooden boats. But it doesn't take many incidents like this one for the practical commercial boat owner to appreciate the advantages of a modern boat built with modern materials.

A lot of good people have been in a lot of tough places. The seiner *Klemtu* found a spectacular spot to wait for high tide. (Martinolich family)

THE *ROYAL PURSUIT*: A NEW SEINER'S HERITAGE

Neil Jensen's new seiner was launched in the spring of 1993, but its story started nearly fifty years earlier when there was still a pilchard fishery on the West Coast. In the late 1940s, Washington State fishers were building some very big pilchard seiners and taking them out off the Olympic Peninsula and south to the mouth of the Columbia River. But within a few years of building the big boats, many of which were over 80 feet long, the pilchard had disappeared. There was a lot of speculation about whether they had been fished out or whether they had just made a temporary change in their migration pattern. When the Canadian government opened the border to allow the import of big US-built seiners in the early 1950s, a bunch of fishers who had built to the demands of the pilchard fishery took the chance to bail out. They either didn't know or didn't care that the Canadian government was encouraging Canadian fishers to buy the big boats for seining pink salmon in the mouth of Juan de Fuca Strait before they reached the American nets.

The Canadian import of these boats affirmed the use of vessel registration as a means of managing the fishery. Eventually this tactic manifested itself in the implementation of a limited entry policy in the BC salmon fishery and the seine fleet was set at about 520 boats. Each boat was measured for length over all and net tonnage. In order to build a new boat it was necessary to retire an old boat of the same size. In the 1980s the naval architects went to work to find ways around the limitations. Where designs had been favouring a dramatic rake in the bow, they went to a nearly vertical bow. Boats were made more box-like in cross-section when compared to the narrow beam and round bottom of the older, wooden boats. Builders took maximum advantage of regulations allowing deductions from the interior volume used in computing tonnage. It has been said that the result was a boat designed by a lawyer and an accountant rather than the traditional team of boatbuilder and fisher.

Boats designed to meet regulations hurt the sensibilities of fishers raised in a tradition of fine-looking boats with good sea-keeping abilities. Neil Jensen was one of these. In 1957, when he was three years old, his dad Emil had the 72-foot wooden *Blue Pacific* built at Mercers Star Shipyards in New Westminster. She was built to a Bill Garden design as a combination seine-longliner and is still noted for her good looks. It was after this boat that Garden went on to design some of the world's great ocean-cruising yachts. The *Blue Pacific* went off to fish the Bering Sea and proved itself to be a fine fishing boat and an excellent sea boat.

THE *ROYAL PURSUIT*

In 1963 Emil switched from wood to steel with the *Royal City*. A contemporary report declared: "Product of a skilled yard and a knowledgeable skipper-owner, the gold-plated *Royal City* is the best fishing vessel of her size it is possible to build. She has a six-way refrigeration system capable of carrying 250,000 pounds of halibut for three weeks; an elaborate electrical system; posh accommodations for nine and the finest electronic and mechanical equipment available. She's the most expensive fish boat ever built in BC, and looks it."

With her gracefully sweeping sheer line, the *Royal City* is still considered one of the classic boats of the coast. Neil Jensen turned nine the year that his dad launched her. His training in the classics was well underway.

Emil Jensen's *Blue Pacific No. 1*, sister ship to Fred Kohse's *Sleep-Robber*, helped form young Neil Jensen's perceptions of a good boat. (PABC 89398)

Emil Jensen's *Royal City*, built in 1963.
(FT/CI 3898-6; VPL 45914)

Young Neil also learned from his dad that to build a boat to your own dream is a proper ambition for any fisher. If the seed was planted by Jensen's first two boats, it was nurtured by the third, the 108-foot *Royal Venture*. It was on this boat that Neil had his real introduction to the adventure of fishing: he went offshore with his dad to fish tuna for two years in the Philippines, Thailand, Singapore and Malaysia on a United Nations program. Returning to BC, Emil was ready to start handing things over to his boys, Neil and Howie. Both thought the *Royal Venture* was too big for the BC fishery, so Howie had master shipwright Ernie Wahl start the *Venturous*. At 70 feet she was the second to last of the big wooden seiners built in BC. (The *Lorinda Lynn*, since renamed *Vicious Fisher*, was launched by the same builder in 1980.) Sadly, Howie became sick and Neil, now turned thirty, took over the boat about the time of his brother's death in 1984.

In the nine years that Neil has fished the *Venturous*, he has developed a bond with the vessel that can be understood only by a fisher who has worked a boat in tough weather and on fisheries as demanding as the West Coast herring roe seine openings. But Neil also wanted the experience of building his own boat. The *Venturous* was too new a boat to be retired from seining if he took the licence off her, and even though it is a good 70-foot length the licence is only for 25 net tons. That's where the *John Todd* came in. She is a big old boat, built in Seattle in 1945, one of the ones destined for the pilchard fishery. Her salmon seine A-licence is for an 82-foot length and 80 net tons on a gross of 122 tons. "I knew that I could build anything I wanted with a licence like that," says Neil. "That's why I bought her."

Neil wanted a proven sea boat and he wanted a good-looking boat.

Knowing that wood is no longer an option for a seine boat, he looked at the alternatives. He wasn't impressed with the aluminum boats that he had seen built to take maximum advantage of their licensed length and tonnage, but he preferred aluminum over steel. A series of fine-lined steel boats of around 75 feet had been built of steel at Allied Shipbuilders in 1978 and 1979, just before the full force of the limited-entry regulations had begun to shape the hulls of seine boats. These boats have a dramatically raked bow flowing into a double-chined hull form that is noted for its sea-keeping abilities. This was the hull design that Neil chose. "But I had her built with fuller bows," he says. "This should give her just a little more lift when she is running into a sea."

Neil selected Al Dawson and Burton Drody's A.B.D. Aluminum Boats & Fabrication in North Vancouver to build the boat, and he went to naval arch-

itect Al McIlwain for the design. The result is a boat 75 feet over all, carrying a 23-foot beam and a 12.5 moulded depth. The dramatic bow rake gives a length on the water line of 65 feet 4 inches, while the 9.6-foot hydraulically operated tilt stern, standard on BC seiners, gives a deck length of nearly 85 feet. The beam is comfortably under the 3:1 ratio that many boats have surpassed in recent years. She has a 66.3 net to 129.27 gross ton ratio. This interesting contrast with the 80:122 registered gross ton ratio of the *John Todd* is largely a result of the modern raised fo'c'sle design

Emil Jensen and Fred Kohse come out on Neil's new seiner *Royal Pursuit* in June 1993 to watch the test sets made in Indian Arm. The new boat, below left, has a tower-style mast. The small boom is used in pumping herring. The tilt stern extends nine feet out past the transom so that when the boat is ballasted down it will reach the water when tilted. Aluminum boats don't require paint, but Neil put a blue and white paint job on his at considerable expense because he feels that it is important for a boat to look good. (AHB)

The power skiff goes into reverse and the seiner goes ahead to start the net off the drum and begin a set. (AHB)

that increased the volume of the hull with deductible crew spaces. "She should load good this one. I didn't want one of those beer cans that you load and they won't move. The hull is proven. The *Angela Lynn*, *Western Investor*, *Snow Drift* and *Bold Performance* are all good sea boats. This is the same hull, I just brought the belly forward and swept up the stern so that hopefully it won't drag the water. With her length and double chine with rolling chocks she'll be a real good sea boat."

Neil feels that the double chine gives him a feel similar to what he is used to on the wooden-hulled *Venturous*: a hull that will roll more easily than a hard-chined hull but will come back well also. It is a hull in which you can feel your boat work in the sea. With all the tonnage that Neil had to work with, it wasn't necessary to reserve the two smaller holds for deductible ballast tanks.

According to estimates used by Al McIlwain, the resulting 3500 cubic feet of hold space will carry 58 pounds of herring for each cubic foot, compared with 64 pounds for a cube of water. This gives the boat a capacity of just over 100 tons of herring. Neil thinks it will probably carry a little more than that, but he stresses that he didn't build it for maximum packing capacity; packers are always available on the herring fishery to take any extra fish. The smaller tanks have also made possible a shaft tunnel to give easy access from the engine room to the shaft, valves and into the lazarette. While this feature may have cost as much as 20 tons of hold capacity, Neil feels it is well worth it.

In the engine room Neil hasn't exercised any restraint, giving the boat a beefy 697-hp Mitsubishi main engine to swing the 69-inch stainless 4-blade prop. A 270-hp Mitsubishi auxiliary runs the main hydraulic pump for the drum seiner, while a pair of Mitsubishi-powered gen sets put out 12 and 25 kw of electric power. The boat has an 18-inch bow and a 15-inch stern thruster.

The new boat, named *Royal Pursuit* in homage to Neil's father's boats, has suitable deluxe accommodations with nice features like a burgundy and bleached oak colour scheme developed by Neil's wife to give a light airy ambience. Each cabin has "cablevision" from the boat's central TV antenna, and Neil's cabin has an extra fold-down bunk for his fifteen-

THE *ROYAL PURSUIT*

The wheelhouse electronics on the West Coast seiner make the bridge on a cruise ship look plain. In addition to the sonar, sounder, radar, loran and a number of radiotelephones, Neil has an electronic chart system. (AHB)

year-old daughter Tara, who fishes with him. The wheelhouse features a yacht-style dashboard display of very comprehensive electronics. It is on the exterior of the boat that Neil's real boat pride shows itself in a major and expensive departure from the normal aluminum seiner. He has painted the whole boat with a shining blue hull and dazzling white house work. "I like a boat to look good," he says, and it does.

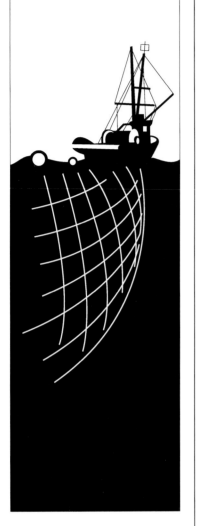

THE
GALLEY BAY

O n Vancouver's False Creek waterfront, where beachcombers once boomed their salvaged logs and squatters' shacks crouched on garbage fill, a trendy pub sits amid rows of pastel townhouses. The valet takes the keys from a pompous young real estate agent who is meeting someone for lunch at Monk McQueen's pub. The name implies a certain style to our young Mercedes driver, and he is more correct in his perception than he probably knows. The reference is to a class of yachts built at the Vancouver yard of George McQueen to designs by the Seattle architect Edwin Monk. It was a collaboration that began in the 1930s and is continued by the two men's sons to this day when one of Ed Monk Jr.'s designs is built at Doug McQueen's modern shipyard on the Fraser River.

Known well beyond the pleasure boat world, the Monk name is carried in the design lineage of many West Coast trollers. Modern Monk designs are limited to big luxury yachts, but in the early days Edwin Monk took whatever customers came through the door. One builder who used a Monk design was Alex North, from Finn Bay up near Lund on the Sunshine Coast. The boat he built to the Monk design was named the *Galley Bay.*

About twelve miles north of Lund, Galley Bay sits at the mouth of Malaspina Inlet on the northern side of the Gifford Peninsula. At the head of the bay is a 180-acre homestead. Like so many others on the coast, it has stood forlorn for years, at once a testament to the dreams and commitment of the settlers who built it and to the frailty of human endeavour.

The homestead at Galley Bay belonged to Axel Hanson, who immigrated to Finn Bay in 1907. In 1912 he took up land at Galley Bay where he remained for the next forty-eight years. His son, Ed, grew up in the bay, but like a lot of other homesteaders' kids, Axel's boy wasn't too excited by the prospect of farming the coast.

Instead, young Ed went fishing, and he was good at it. Good enough that by 1946 he could go to Alex North down at Finn Bay and order a brand new boat. "I sent down to Seattle and got an Edwin Monk Design," Ed recalled in 1989. With an inside beam of

The *Galley Bay* at the Campbell River
fuel dock. (AHB)

12 feet 2 inches on her 41-foot length, the new troller was considered big for her day—too big, some said. Ed Hanson named her the *Galley Bay* after his childhood home. He put in one of the new 110-hp Chrysler Crown gas engines and went fishing. She was a popular design, and Ed remembers that Alex North built six of the boats altogether. In addition to the *Galley Bay* (1946), they were the *Norlite* (1947), *Lor-Dell II* (1948), *Wanderer No. 2* (1949), *Adolfina* and *Finn Bay*.

If doubters thought the new generation of trollers were too big, they soon had reason to swallow their criticism. In 1948 the warm water currents brought the tuna in close to the Queen Charlotte Islands, and boats with the packing capacity made good money. Ed Hanson was able to take the *Galley Bay* 60 miles offshore to the best fishing. He fished tuna again in 1960 but this time he had to go 120 miles off the Charlottes. Just as his dad kept working at the Galley Bay homestead year after year, Ed kept fishing his *Galley Bay*. In 1960 he replaced the gas engine with a Gardner 6LW. Other than that, the boat that Ed Hanson took out salmon fishing in the 1980s was pretty much the same boat launched in Finn Bay.

As Ed contemplated retiring, and selling his boat, he reflected on the fate of his dad's farm. "He stayed there until 1960 and sold it to some Americans. They made a hippie place out of it. Dad's beautiful yard and vegetable garden was used for growing 'pot'. Dad had the 180 acres and eleven-room house in perfect shape. Last time I saw it they had even burned the picket fence. Those were all hand-split cedar pickets. They were too lazy to even cut firewood. It just breaks my heart to see how things have gone to hell!"

Ed Hanson. (AHB)

Ed never did quite get the heart to sell his boat. Maybe he worried that whoever bought it would treat it as badly as the buyers of the old homestead had treated it. When he died in 1990, his son sold the boat to Ralph Miller over in Port Alberni. Ralph has fished since 1963. He owned a big freezer troller that froze the salmon at sea, and he stayed out until the hold was full. But this required two deckhands to help with the glazing and generally keep up with the work. It also meant that the freezer engine had to be running twenty-four hours a day, seven days a week. By buying the *Galley Bay*, Ralph went back to icing his fish. It means having to deliver the fish more often, but he and his wife Ruby are able to fish the boat in a more relaxed style.

Ralph prefers to fish the northwest coast of Vancouver Island, but he travels to wherever he thinks the fish will be. In the summer of 1992 he travelled to Prince Rupert. "A lot of people came down to the boat and told me how they knew the boat from when Ed had it. Some had decked on the boat with Ed and everyone spoke very well of him. Some people asked me if I would change the name from *Galley Bay*, but I have way too much respect for Ed Hanson to do that. The boat and Ed grew old together so I have had to do some work to it, but she is a real good sea boat."

If Ed worried that the boat he named for his parents' homestead might pass into uncaring hands, he would be pleased with her new home.

AN IMMIGRANT SUCCESS STORY

Like most immigrant fishers, Tram Van Tam arrived in Canada from his native Vietnam with little more than a few dollars in his pocket, a dream and a will to work hard. He went first to Prince George, where he got a job in a sawmill, while his wife, Trinh My Ha, and baby daughter stayed in an unfurnished apartment. They slept and ate on the floor until they saved enough for a little furniture. When Tam was laid off from the sawmill, they moved to Vancouver. Tam worked at various jobs, including a year as a gardener, while he saved the down payment for his first fishing boat.

In Vietnam Tam had started fishing when he was twelve years old. He served a three-year in-breaker period in the galley, then at fifteen he was promoted to the deck and better pay. He spent another ten years as a deckhand learning to fish the long gillnet and to trawl for shrimp.

Trinh My Ha's mother and father were fish brokers who owned three large boats. They had seven daughters and no sons. Although there are some women in the Vietnamese fishing industry, My Ha did not fish, so when she married Tram Van Tam he became skipper of one of the three boats.

At 105 feet it was a big boat, but the gear was all hauled by hand. They sometimes caught tuna in their gillnets. At times they trawled shrimp using a net rigged with wooden doors rather than the beam that Tam uses on his Canadian shrimp trawl. They also seined for fish that looked like small herring. At first the nine-man crew pursed the seine by hand, but later a mechanical winch was added.

Tam skippered this boat for three years. But with the American defeat in South Vietnam, conditions were changing for his family. Finally, in the spring of 1980, they made the decision to leave their homeland. There was no shortage of passengers: the fish boat was loaded with 123 people for the three-day trip to Thailand, including Tam, his wife, their six-month-old child Tram Poa Loan and a number of other family members. When they arrived in Thailand their boat was seized by the government and they spent six months waiting

SUCCESS STORY

Opposite: In the past the shrimp fishery filled in the time for a few fishers over the winter. Vietnamese immigrants have helped develop it into a major off-season fishery. This boat carries the beam that is used to keep the mouth of the shrimp trawl open at the same time as it moves across the bottom stirring the shrimp up from the mud. (AHB)

Left: Tram Van Tam, with the hat, on board a fishing vessel in Vietnam. (Tram family)

for the opportunity to move on to Canada.

Tam was an experienced twenty-nine-year-old fishing captain when he arrived in Canada, yet the Canadian government sent him several hundred miles inland to Prince George. Years passed and three more children were born: the twins, Linda and Helen, and daughter Amy Tram. By 1984 Tam was able to buy his first boat. It was an older wooden boat that he fished for two seasons. In spite of persistent rumours to the contrary, Vietnamese newcomers did not receive any government assistance to buy their boats. Tam saved what he could and borrowed some more from friends for the down payment, and borrowed the balance from the bank.

After the 1985 season he sold the wooden boat and bought a bigger fibreglass boat with his brother. A year later he sold his share to his brother and put a down payment on his own boat. In February 1987, he bought the 39-foot *Minato*. It is a modern gillnetter which he uses to fish shrimp or rigs for longlining. To get the most from his fish he sells them at the new public sales dock on the Steveston waterfront when he can. When I visited the family in 1992, the kids were old enough to help a little and the family seemed to have become a typical immigrant success story.

"I go fishing every week," Tam said. "In summer, when the gillnetting closes for the weekend, I take off the salmon net and put the shrimp net on. I work hard to pay off the bank and my friends." By November when the salmon season ends, he is fishing shrimp up around Gibsons and coming down to Steveston for the weekend where his wife My Ha sells the week's catch from the government dock.

Trinh Thi An stands with her daughter Trinh My Ha and her son-in-law Tram Van Tam and the family's four children in 1987. (AHB)

A pair of seiners await a herring opening with their packer, the Canadian Fishing Company's *Cape Scott* (built 1943 in Antioch, CA), ex-*Coastal Trader II*. It was built for the US military. (IW)

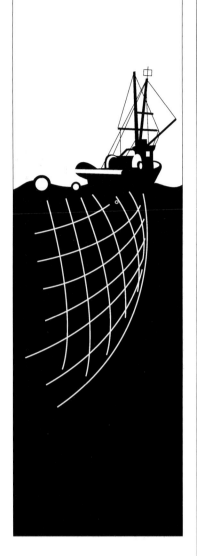

WOMEN IN THE INDUSTRY

Like farming, the commercial fishing industry has always been family oriented. Since the earliest days of the First Nations fisheries, women have played an important role, usually at the processing end, whether it was cleaning and preserving the fish or working a line in one of the coastal canneries. Almost always in fishing, women have had the double role of worker and mother, and almost always as someone's wife. In the old records of the ship's registry when a man owned a boat his occupation was listed as "businessman" or "fisherman" while a woman who owned a boat was listed as "wife of . . ." or "widow of . . . ." In practice, it was often as "widow of . . ." that a woman turned to the only work that she knew and took the family boat out to the grounds.

The financial management of the family business has been a major, if unheralded, role that many women have played in the fishing industry. Owning and operating a fishing boat is a business, one that requires government form-filling and the careful bookeeping and cost accounting of any small business. Invariably done by women, it is work that if left undone would have seen many more good fishers lose boats to the banks—not because of poor fishing, but because of poor management.

In all of the fishing stories that I have heard I know of only one woman seine skipper. Pat Roberts Piatocka ran the little boat *Ruston* about 1963. The boat, the poorest in Nelson Brothers' large fleet, was in terrible shape. Pat managed to nurse it and a crew through the season, but she didn't bother trying again. On the gillnetters, trollers and packers, on the other hand, there have now been enough successful woman skippers that it is no longer a novelty.

In a letter to the editor of a commercial fishing journal in 1990, young Syreeta Caron wrote wondering if there were any other girls fishing with their dads. In response, fifteen-year-old Joeline Silvey, daughter of Chris Silvey, wrote: "Well I'm happy to say that I've been fishing with my dad for four years on the *Roona* and the *Pacific Comber*. I have been on and around boats since I was born, but every summer I have yet to see another girl deckhand so I know how Syreeta feels. But have hope, there are a precious few of us out there and we're secretly taking over."

In her optimism, Joeline may have been a little ahead of herself, but there is no question that a great many more women are taking command on the grounds. There has long been a solid core of female deckhands and owner-skippers. This is particularly true in the trolling fleet where women have a history of successful participation. Joan Lemmers,

Pat Roberts Piatocka, skipper on the *Ruston*, in 1963. (AHB)

for example, has proven herself as a fisher with the well-known *K'san II* and has also become an outspoken environmentalist, speaking wherever she can about the commercial fisher's responsibility to the waters and the fish.

Among the gillnet fleet there have been a number of successful women. Nancy Marshal of Ladner began her fishing career by working as a social worker by day and going out with friends for evening drifts on the Fraser. She caught the sockeye fever and worked her way up through a series of boats to eventually buy the 50-foot Frostad-built packer *Hey Dad*. She used this boat to buy fish from her old gillnet comrades before selling it to move to a shore manager's job. By the spring of 1993 Nancy had had enough of being stuck ashore and was planning to take the packer *Alpaca* out for the summer salmon season. She had bought a 25 percent share in the boat a few years earlier. It's a beautiful boat with a romantic history: it was named by its first owner, Al Capone, who had it built at Shelburne, Nova Scotia, in 1927 for his East Coast rumrunning operations.

Nancy explains that there are some fisheries, such as black cod or Bering sea halibut, that she would rather not try as they are known for being particularly tough. But she also points out that much of this is perception. She recalled sitting in Ucluelet with another woman, Dawn Vanstone, in 1977. They both wanted to go out on the herring gillnet fishery, which at the time was noted for being physically demanding. They watched as two fishermen struggled to carry a 45-gallon drum of gas down the dock. Realizing that a full drum weighed about 450 pounds, they were just deciding that they weren't strong enough when the men dropped the drum and the two women saw it bounce. It was empty. They looked at each other and both said, "Next year we go herring fishing!" And they did.

Among the most successful women gillnetters have been the Sowdens of Sointula.

Nancy Marshal on her packer *Hey Dad* in 1987. She later sold the boat and tried to come ashore, but in 1993 sockeye fever got the best of her and she went back out. (AHB)

Linda Sowden, who fishes the modern aluminum gillnetter *Wolf Point*, learned the art from her mother and father, Christine and Robert. Christine's side of the family had the longest involvement with the industry. Her father's mother began gillnetting on the Skagit River in Washington State when her fisher husband died about 1900. This was in the days of row skiffs and long skirts, but even then it was possible for a woman to handle a gillnetter. Christine's father grew up fishing and took up the life in his turn, "with a bit of time out for rumrunning in the Gulf Islands

Linda and Christine Sowden in Sointula, the fourth generation of women fishing the tides. They are standing in front of a herring punt with a beater bar for shaking the herring out of the gillnet.
(AHB)

where he met my mother," says Christine. Christine's father died in Alaska and she was raised with her mother's family, loggers in the Gulf Islands on Prevost Island. When she was nineteen years old, in the late 1940s, Christine bought an old boat and began longlining dogfish. She was hampered a bit by her mother who insisted on coming along to "show her how," until one day when "I put an anchor right on the stern and she baited a bunch of hooks and had it all coiled up. She gave the anchor a big heave and one of the hooks came around and caught her in the seat." After Christine cut the hook out, her mother finally recognized her as skipper.

When the family moved north to Adams River and Port Harvey, Christine bought a bigger gillnet boat with a 2-cylinder engine. She planned to take this boat north, but before she did, she met and married Robert Sowden. Robert was a logger, but he also had a small troller. "That was when the government gave all of the country round about to the big companies and put all the little fellows out of business," Christine remembers, "so we decided to go fishing."

Robert's troller was not as good as Christine's gill-netter, the *Linda*, so they sold it and went gillnetting. That first year, 1953, the young couple learned to gillnet in Rivers Inlet and out in Johnstone Strait. When their first daughter was born they reversed the common practice of naming the family boat after the daughter and instead named the daughter after the family boat—Linda. They continued to gillnet salmon but moved, first to Sointula and then to Victoria when Linda and her younger brother Mike needed to attend school. They spent nine years in Victoria, but returned to their Sointula home when the kids finished school. By this time it was the early 1970s and Linda was pushing to take over the family boat, which was now the Yamanaka-built *Ripple Rock*. In 1973 the parents had the *Dawn Treader* built and Linda and her brother took over the *Ripple Rock*. A year later she had her present boat, the *Wolf Point*, built at Shore Boats on the Fraser River. The family fleet of aluminum gillnetters was completed when Mike had a new boat built in the spring of 1987.

Christine explains the family's success: "We pay attention to what we're doing when we're doing it. So many will drink all weekend, others get all kinds of funny ideas about how and where they should fish—fables. We just go and fish and think, 'What is the fish thinking about?' We don't have a horrific number of nets. We have usually worked with just two: a sockeye net and a fall net [for chum salmon]. They are heavier nets than most."

The family fishes the Nass and Skeena Rivers. In the fall Mike and Linda go over to the Queen Charlotte Islands for dog salmon. Robert died in 1988 and Christine went on to fish another season, but when she found herself wrestling with a big dog salmon that came thrashing over the stern of the gillnetter, she decided there wasn't much to be gained and sold the boat. When asked if they fish any differently because they are women, Christine responds, "No, you just charge in there and fish." Linda commented: "My brother usually beats me because he's got more muscles than me. When I get tired he's still going.

"But there have been a few occasions when I beat him," she adds with the satisfied smile of a competitive fisher.

THE ZEN OF SEINING

T he sea has little patience for fishers who attempt to control it. When at sea the duality of boat, or engine, and self must not dominate one's consciousness. The skipper, crew, boat and environment must be in harmony if success is to be maintained over the long haul.

This is not to say that life on a seine boat is one of idyllic harmony. Fred Kohse speaks of "gumption" to define that determination to excel against all odds. In his book *Zen and the Art of Motorcycle Maintenance*, Robert Pirsig elaborates on the concept, speaking of "gumption traps" as those things which drain gumption. Fishers, with fitting irony, call these "jackpots," meaning the opposite of the more conventional poker winnings jackpot.

The hull of a seine boat defines a very precise world. When problems arise, they will be solved with the finite resources and expertise of that boat and its crew, or else fishing time or even the boat itself will be lost while waiting for help. Every skipper gets into jackpots. The test of the good skipper is how calmly and quietly—how harmoniously—he gets out of them.

Herb Assu was a skipper noted for his calm and absolute harmony with the sea: its tides, its weather, and its fish. Herb was a third-generation seine boat skipper, in the tradition of his father Dan and his great-grand-father Billy, who once told his crew that he could spot more salmon with his one glass eye than they could with all their good eyes. For many years prior to 1980, Herb was the owner and skipper of the *Departure Bay No. 3*, a beautifully proportioned 58-foot seiner. It was built in 1927 by Jisaburo Sawada at the Ode Camp boatyard on Newcastle Island near Nanaimo and was used by the Hamagami family in their herring fishery. Once, when one of Herb Assu's crew,

THE ZEN OF SEINING

who had been to university, suggested keeping a log book to record tide, water temperature and other observations for each set of the net, Herb's only reaction was to tap his temple with his forefinger.

In lower Johnstone Strait, Herb's home fishing grounds, a common type of set is one in which one end of the net is tied to a "peg," or tree, on shore and the other end is towed against the flood tide. The law requires that the net be held open for no more than twenty minutes. Then the seine boat circles back into shore where the tie-up man releases the end of the net. From this point in the set, while the net is pursed up and pulled back aboard, the boat is effectively without propulsion because the net would tangle in the propeller if it were running. Depending on wind, tide and location, there is a somewhat predictable length of time before the boat will hit the rocky cliffs which make up most of the shoreline in this part of the country.

One day, when Herb was drum seining with the *Departure Bay No. 3*, he had just started to drum his net aboard when a shaft key on a gear that is essential to the recovery of the net worked its way along the groove in the shaft and dropped overboard into about 30 fathoms of water. Everything stopped. A strong west wind combined with the flood tide to turn the boat at the same time as it moved it steadily toward a rocky point. There were no boats near for a tow. Herb asked for a wrench, disassembled the offending gear, and laid it on the hatch cover. Then, with $20,000 worth of uninsured net wrapped around an underinsured $100,000 boat, Herb sat

Oscar Lewis's crew pitch fish on to the packer in 1946. The light bulb suspended over the hatch helps the crew identify the catch and avoid the scorn of the packer men who watch for lower value fish being passed off as higher value. (FT)

Opposite: The *Departure Bay No. 3* rides at anchor, waiting for the tide in lower Johnstone Strait in the 1960s. (AHB)

In the early years members of the seine fleet had a special position in their communities. A magistrate in Campbell River in the 1950s routinely granted ten-dollar bail to any seine boat crewman who got in trouble on a Saturday night because it was known that he was needed on the boat the next day. In this photo from 1946, Oscar Lewis's Cape Mudge crew brails salmon aboard the *Departure Bay No. 5* while Robert Clifton Sr. of Comox works with his crew to haul back the seine on the *Kwatsu*. (FT)

VANCOUVER B.C.

Herb Assu was truly at home on the water. He felt more of the tide through his feet planted firmly on the deck than most university kids could understand in twenty years of theory. (Assu Family)

The seiner *Eskimo* closes a set in Johnstone Strait in 1946. (FT)

down on the hatch and looked at the gear. The rest of the crew did the same. The boat drifted. Someone was sent to the engine room to see if a spare key might just be in that jar of odds and ends. Minutes ticked past as the shore loomed closer. No spare key. Everyone looked at the little quarter-inch groove on the inside of the gear and meditated. They could hear the breakers on the beach when Herb sent someone below again, this time for a file and a hammer. When the file arrived he laid its quarter-inch-square handle end in the gear's groove. A perfect fit! Fifteen minutes of fast work later the file end had been broken off with the hammer, the gear reassembled, the net straightened out and drummed back aboard. When Herb kicked the main clutch into reverse, barely ten feet of water separated the bow of his boat from the rocks. Harmony.

All of this contrasts with those skippers, often young, who believe the solution to any problem can be found in the radiotelephone. Today this is often true, but the radiophone call advertises the skipper's jackpot so widely that no self-respecting skipper will use it in any but the most dire emergency. The young skipper who makes a habit of phoning his dad or uncle to ask, "How's the tide over there?" and asks for a tow every time he gets tangled in his own net, will not be accepted into the fraternity until he learns a little restraint.

On the weekend, when crews get together, talk will occasionally turn to a comparison of skippers. The skipper who is a driver, and uses a bigger diameter beach line in order to set in stronger tides, is permitted to yell at his crew from time to time because he catches a lot of fish. But the skipper who can't read the tide and is constantly breaking beachlines and getting big "roll-ups" in the net, and then screams out his frustrations at his crew, will have trouble getting good people.

Lack of gumption leads to panic. During the 1970s the price of roe herring went over $2000 per ton. This led to a lot of overloading of seine boats during the season of winter storms. As a result, a lot of boats sank and too many men drowned. It is hard to meditate when there are fortunes to be made. A fever of excitement grips men. It was in this season that the engineer of one large seine boat went into the engine room to check on things while the boat was travelling in with several other boats in a strong sea. Several of the crew who were sitting in the galley were surprised to see him emerge from the engine room screaming that it was flooding. At the same time he pulled on his survival suit, ran out on deck, and jumped overboard. Fortunately, one of the other boats picked him up. The seine boat did sink, but the rest of the crew were able to go directly onto the rescue boat without such loss of dignity.

Herb Assu's uncle, Jim Quatel, pulled a line of Spanish corks on the seiner *Frank Ellis* at Namu in 1945. (NFB/1971-271, Item 16888)

Herb Assu never lost a boat, but there were times when he might have worried. On one such occasion he awoke at two in the morning because his boat, which had been riding at anchor, had developed a barely perceptible list. The anchor had dragged, allowing the stern to settle on a ledge just out from a gravel beach. Herb had anchored in this spot for years and knew that not only was the boat in danger but the potential for considerable embarrassment existed as well. The tide was dropping fast. Herb woke the crew and started the engine. The anchor was winched aboard, then lowered into the skiff. Herb directed the skiffman to row it out as far as possible so that he could try to pull the boat off with the anchor winch. On the first try the anchor just dragged all the way back to the boat. By now the boat was listing so that the top of the mast was over the gunwale. Try again. This time the anchor got a firm purchase on the bottom, but now the weight of the boat was too much for the anchor winch to pull. A powerful spotlight beam broke the darkness. A passing tugboat was willing to help—for a price to be settled after the boat was off the beach. The skiffman was sent out to the tug with the message: "Thanks, but no thanks." By this time the starboard deck was awash and water was creeping up the hatch combing.

The skipper's wife and small child were put ashore and at the same time the skipper directed the end of a spare purse line to be tied to a big cedar some distance up from the shore. The other end he passed through the starboard pursing davit and around the main winch head just behind the cabin. As he put strain on the line the already listing boat tipped crazily, and the water came up to the galley door, but then the powerful main winch swung the boat parallel to the shore until her belly pulled up tight against the underwater ledge. By now the grey dawn had reddened over the Coast Mountains behind Thurlow Island. The crew settled down to wait out the low-water slack. They had anchored here to be in position for

The *Izumi II*, built for the herring salteries, fishes salmon in Johnstone Strait off Ripple Point in 1955. The power block has just been introduced, but it is either not working or not trusted as it sits unused on the cabin top while the crew pulls by hand. (FT)

a morning set during that time, but would have to wait until high water in the morning floated the boat free. In the meantime, with strong ropes holding her to the shore, all danger was past.

Herb was standing in the stern as someone turned on the radio-phone. Bull Harbour Coast Guard radio boomed in to tell all boats that the fishing vessel *Departure Bay No. 3* was in need of assistance below Camp Point. Now the man who had been absolutely calm for the past three hours of "gumption traps" sprinted his not-inconsiderable bulk across the deck and forward to the pilot house to get on the phone and assure the Coast Guard that the *Departure Bay No. 3* was just fine, thank you. There are some things that can cause panic in even the most experienced skipper.

These are the stories of the exceptions, the times when a wrong decision could mean a day's fishing lost, a boat sunk, or even a whole crew drowned. The day-to-day skippering of a seine boat is not so spectacular. However, if fish are to be caught, it requires that same calm assessment of many factors that present themselves in an infinite variety. When Herb Assu stood behind the dodger on the cabin top of the *Departure Bay No. 3*, he felt the tidal currents through the boat at the same time as he observed the faintest line and rip on the surface that told him if the tide had really straightened out. Anyone can look at the kelp heads to see which way the surface tide is running, but only a master will know from surface observations exactly what is going on 50 feet under the surface. And tide is only a part of the picture. The amount of daylight is also important. Set too early in the morning and you may get hundreds of little rock cod gilled in your net—a big picking job. Time of year, amount of wind, and a host of other considerations add up to what can only be called harmony. It is in this that the seine skipper, the fly-fisherman and the Zen disciple become one.

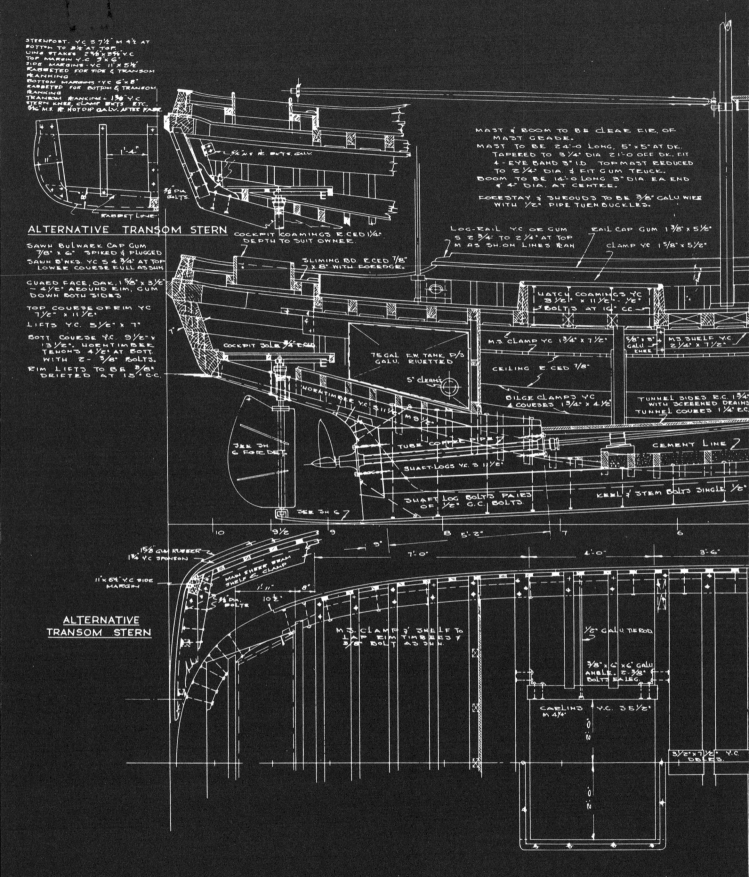